THE ECONOMICS OF UNCERTAINTY

PRINCETON STUDIES
IN MATHEMATICAL ECONOMICS
EDITED BY OSKAR MORGENSTERN, HAROLD W. KUHN,
AND DAVID GALE

THE ECONOMICS OF

Uncertainty

BY KARL HENRIK BORCH

Princeton, New Jersey
Princeton University Press

Publication of this book has been aided by the
Whitney Darrow Publication Reserve Fund of
Princeton University Press

Printed in the United States of America
by Princeton University Press

Second Printing, 1969
Third Printing, 1972

Preface

At the end of World War II, Field-Marshal Wavell published an anthology of poems, which he called *Other Men's Flowers*. This would have been a suitable title for this book if it did not create the impression that economists do not take themselves seriously.

To prepare this book, I have freely picked the results of other people when they seemed beautiful to me, and when I could fit them into an arrangement which I hoped would be attractive. Not being Japanese, I will not claim that flower arrangement is an art, and I find it difficult to argue that it is useful. I hope, nevertheless, that this book will be useful to students of economics. I hope the book will enable them to see that a number of recent results from very different fields of economics—and from other social sciences—really constitute an entity. These results are impressive and often of a startling nature. They seem to fascinate many young economists and, to use another botanical metaphor, make them see the trees instead of the wood. The purpose of this book is to try to convince the potential reader that, although the trees may be extremely interesting, it is the beautiful wood which is really important. We live in it.

It is impossible to prepare a book of this kind, without incurring a heavy debt to a great number of colleagues. The debt is obvious where I quote the work of others. The debt is real, but less obvious, in many other places. There is an oral tradition in the teaching of the subject of this book. This tradition consists of a number of examples, counter-examples, and paradoxes which have been created by ingenious teachers, but which too often are taken as part of the common knowledge. I have drawn heavily on this tradition, and I find it impossible to mention by name all those who should be mentioned and to acknowledge my particular debt to them.

I must, however, express my particular thanks to Professor Oskar Morgenstern. Without his constant prodding and encouragement over a number of years, this book would never have been written. It is also his work which led to the breakthrough which opened the way to the results which I have tried to survey and bring together in this volume.

The others who had deserved a special expression of thanks, will, I believe, acknowledge their own debt to Professor Morgenstern, and I hope they will not be offended when I single him out in this way.

The book has found its present form after a number of years, and is the outcome of lectures I have given to graduate students at The Norwegian School of Economics and Business Administration in Bergen, The Institute of Advanced Studies in Vienna, and The Graduate School of Business Administration of the University of California in Los Angeles. I probably owe a greater debt than I realize to these students. Their questions and comments have certainly helped to clear up a number of points which I fear must have been obscure when I first presented them.

Karl Borch

Bergen, August 1966

Contents

THE ECONOMICS OF UNCERTAINTY

Chapter I

The Economics of Uncertainty

1.1. The theory of economics consists to a large extent of theories about how people make decisions. When we set out to study the role which uncertainty plays in economics, it seems natural to begin with the subject usually known as *decision-making under uncertainty*. This is a familiar term. It has been used as the title for a number of books and articles, and it has also been used to describe university courses; it can be found as a separate heading in the classification system of libraries. It might, therefore, have been appropriate to call this book "Decision Making under Uncertainty—with Application to Economic Problems." The more general title, *The Economics of Uncertainty* is, however, preferable for a number of reasons, which we shall discuss in some detail in this introductory chapter.

1.2. In the simpler decision problems—in business or in other fields—we take the world as given, at least in a probabilistic sense, and try to find the "best" or the "optimal" decision.

The classical example is that we take prices, or their probability distributions, as given and decide how we should spend our money in order to obtain the maximum satisfaction. If we are in business as a producer, the problem is to figure out how much we should produce in order to maximize our profits, or expected profits.

The inevitable question is then, "Prices and other things are given by whom?" If they are given by God or the government, there is little we can do about it. We are on the defensive, but we know where we stand. We can figure out our best course of action—which eventually may turn out to be quite profitable.

For instance, let us assume that a farmer can plant either potatoes or wheat in the spring. Let us further assume that if there is a dry summer, potatoes will be best, but wheat will give the best crop if the summer is wet. If we complete the description of this situation, we will get a typical example of a decision problem under uncertainty. In order to get this complete model, we may have to specify elements like the probability distribution of rainfall in inches and the relationship between rainfall and the yields of potatoes and wheat.

The solution to this problem may be that the farmer should plant wheat on 40% of his land and potatoes on the rest—or in any case that he should make a decision which can be stated in this form. The implication is that with this solution he will do reasonably well no matter what happens, and he will do quite well if the most likely "state of the world" should actually materialize.

1.3. If prices and other elements in our problem are not given by some higher authority, they may be the outcome, or the result, of decisions made by other persons who are in a situation similar to our own, If this is the case, the problem will change radically.

Let us, for instance, assume that all meteorological forecasts agree that it will be a dry summer. This should mean that every sensible farmer would plant potatoes on all his land. If they all do this, the price of potatoes may fall dramatically, and the crazy—or shrewd—farmer who planted wheat will profit from a general shortage of wheat. This farmer may have been out of touch with the world, but he may also have outguessed the world by thinking of prices, which obviously are influenced by the decisions made by other farmers.

1.4. As another example, let us assume that we have some money to invest in the stock market. We can study the prospects of the various companies and the prices of their stock and may then be able to select one particular stock as the "best buy" in the market. We can use advanced mathematics and sophisticated arguments to reach this decision, and we may consider ourselves very clever.

However, if we buy the stock in question, there is necessarily a seller who thinks that at the present time and the present price it is right to sell the stock which we consider the best buy. If this seller is just as intelligent and smart as we are, it may be useful to think twice.

1.5. The point we want to make with these two examples, is that there are two types of decision problems, and the difference between them is important, and in some cases fundamental.

If our decision problem is what we can call a *game against nature*, we may have to take the problem as given. We may, of course, try to find out more about the laws of nature in order to reduce the uncertainty and make our decision easier. This will, however, lead to a completely new set of problems, which we shall not consider in this book.

If, on the other hand, we have to make a decision in a *social context*, the problem may not be "given" in the same sense. The *data* of the problem may be determined by the decisions made by other persons who are in a situation similar to our own. If we assume that these people are intelligent, and think in the same way as we do ourselves, the problem becomes more difficult. However, this also provides us with important information, which we can use as we reason our way to the solution of our own problem.

Let us, for instance, assume that our farmer decides to spray his crop with a certain insecticide early in the season. This may give a bigger harvest, measured in tons or bushels, regardless of whether other farmers around the country have sprayed or not. The spraying may also give a better harvest, measured in dollars, but this may depend on what other farmers have done.

1.6. To illustrate the point, we shall consider an example discussed by Howard in his book on dynamic programming ([1] pp. 54–59). Howard assumes that the following four elements are given:

(i) $C(t)$ = the cost of buying a car of age t;
(ii) $T(t)$ = the trade-in value of a car of age t;
(iii) $E(t)$ = the expected cost of operating a car of age t until it reaches age $t + 1$;
(iv) $P(t)$ = the probability that a car of age t will survive, i.e., reach the age $t + 1$.

He then seeks to determine the trade-in policy which will minimize expected transportation costs. It is clear that a "policy" in this context must be a rule which tells us what to do if we own a car of age t. We can either keep the car for another period, or we can trade it for another car of age $s = s(t)$. This means that a policy is defined by a function $s(t)$ defined for all t. A more precise mathematical formulation would be to say that a policy is a *mapping* of the set of ages on itself.

1.7. Let now $V(t, s)$ be expected transportation cost per period if we own a car of age t and have adopted the policy defined by the function $s(t)$. It is easy to see that $V(t, s)$ must satisfy the difference equation

$$V(t, s) = C(s) - T(t) + E(s) + P(s)V(s + 1, s).$$

The first two terms give the cost of trading our present car against one of age s. The third term gives our expected transportation cost in the next period. After that period we will, with probability $P(s)$, be the owners of a car of age $s + 1$, and this is expressed by the fourth term.

If we have $t = s(t)$, i.e., do not trade for some value of t,

$$V(t, s) = E(t) + P(t)V(t + 1, s).$$

It is a formidable mathematical problem to solve this equation and then determine the optimal policy. In fact it seems practically impossible to solve the problem by this direct approach. Howard's important contribution consisted in developing a relatively simple algorithm which enables us to compute the optimal policy in a few steps. His book is in many ways a mathematician's delight, but his formulation of the problems in general differ from the formulations we shall find natural to discuss in this book.

1.8. In a numerical example, Howard assumes that the age of a car can be described by an integral number of three-month periods, and he assumes that $0 \leq t \leq 40$, i.e., $P(40) = 0$. This means that a car which is 10 years old is certain to break down beyond repair within 3 months. He then specifies the four functions in 1.6, in an apparently realistic manner, and finds that the following policy is optimal:

(1) If you have a car which is more than 6 months old but less than $6\frac{1}{2}$ years old, keep it.
(2) If you have a car of any other age, trade it in for a 3-year-old car.

This policy is defined by the function

$$s(t) = t \quad \text{for} \quad 2 < t < 26$$
$$s(t) = 12 \quad \text{for} \quad 0 \leq t \leq 2$$
$$\text{and } 26 \leq t \leq 40.$$

Economically this is nonsense. What would happen to the automobile industry if rational consumers only wanted to buy 3-year-old cars? Howard is of course well aware of this, and he makes it quite clear that he discusses the example only to illustrate the application of his algorithm.

In this book we shall discuss a number of problems similar to this one. We shall, however, not discuss algorithms or computational procedures in any detail. In general we shall study the problems from two points of view:

(i) We shall examine the *basic assumptions* which must be fulfilled in order to justify the application of particular techniques.
(ii) We shall discuss *reformulations* of the problem which may give it more economic meaning.

1.9. To illustrate the first point, let us note that it is clear that the problem we have discussed may exist in real life, if we can assume that car owners fall into three classes:

(i) People who enjoy driving new cars even if this implies higher transportation cost than strictly necessary.
(ii) "Rational" people, whose sole objective is to minimize their expected transportation cost.
(iii) People who drive old cars because they lack the capital (or credit) which is necessary in order to switch to a newer and more economical car.

It is obvious that we can think of a number of other and quite reasonable assumptions, which would make economic sense out of Howard's model.

1.10. To illustrate the second point, let us see how we can reformulate the problem in order to make it more interesting, at least to an economist, if not to a mathematician.

The elements which it is natural to take as given are $E(t)$ and $P(t)$, i.e., the operating cost and the survival probability, which we can assume are determined by purely technical conditions.

In economics it is not always natural to take prices as given. A real economic theory should explain how the prices are determined within the system. In Howard's automobile market there will be a scramble for cars which are just three years old. The lucky owners of these cars may then raise their selling price—after all they do not want to trade away their good old car if this leads to higher transportation cost. This means that the three-year-old cars will lose their conspicuous position as the obvious best trade in the market.

Let us now assume that the trade-in margin $C(t) - T(t)$ is given, for instance that it is determined by the car dealer's expenses. We can then try to find a set of prices, i.e., a function $C(t)$, such that no trade-in will reduce our expected transportation cost. This is a meaningful—in fact classical—economic problem. If such a set of prices exists and these prices are accepted as the "market prices," nobody can gain by trading; i.e., the market will be in *equilibrium*. In the following chapters we shall discuss many examples of this kind of *static equilibrium* and determine the corresponding prices.

1.11. Static equilibrium is a concept of considerable theoretical importance, but it is often rather remote from the economic situations which we study in real life. Even in Howard's simple model things happen as time goes by. Cars grow older, and some of them go out of operation. We must also assume that new cars are brought to the market according to some rule or production schedule. To master this situation we must construct a *dynamic* model, and in order to describe a model of this kind, we will usually have to specify far more elements than in the simple static model.

If, for instance, we specify the initial age distribution of cars, we can determine the expected replacement demand for new cars in each successive period. If in addition we introduce some "elasticity" into the model, we may determine the number of new cars which can be sold at a given price.

However, the price is not usually given by some higher authority, so this formulation of the problem will give a functional relation between price and the number of cars which can be sold at that price. If this relation is known to the producer of automobiles, it is natural to assume that he will use it to decide on the number of cars he should bring on the market in order to achieve his objective, which may be to maximize profits—in the short (or in the long) run.

It is also natural to assume that the price of secondhand cars will depend on the number of new cars offered in the market. This assumption implies, however, that the decision problem of the car buyer depends on the decision made by the producer; i.e., the buyer does not really know what his problem is unless he knows the decision made by the producer, and vice versa.

1.12. The model we have outlined in the preceding section illustrates something essential in economics and other social sciences. In economics there will, except in degenerate Robinson Crusoe examples, be two or more parties to any transaction. If there is no compulsion, all parties must think that they benefit from the transactions which take place. In economics we usually assume that people act to advance their own interests, as they see them. If we want to understand what happens in the economy, we must look at all parties to the transactions. This holds also if we just want to "mind our own business" and think only of advancing our own interests. This again means that we cannot solve each decision problem separately. We have to solve the decision problems of all parties simultaneously.

The best and most clear-cut example of such interactions is probably the *two-person zero-sum game*, which we shall discuss in some detail in Chapter IX. In this model we can figure out exactly what our opponent will do just by assuming that he is intelligent, or "acts rationally," i.e., as we would do in his situation.

1.13. We have used the word *transaction*, and this may be a useful key word. In engineering and natural sciences, a very popular word is *process*. The laws governing the process are usually taken as given, and the decision or management problem is to "control" or steer the process so that it runs in the way we like best or, to be specific, which is most profitable.

To illustrate the point, let us again look at the stock market. We can, of course, look at stock prices as a "process" and try to find some way of making money by buying and selling as the process develops. However, we may, or we may not, gain something by thinking of the transactions behind the process. If a stock has a *market price*, this must mean that there is some kind of "equilibrium" between those who want to sell and those who want to buy at that price.

1.14. We have outlined one possible approach to our subject, which we can summarize as follows: Starting from the more classical decision problems with uncertainty, we can look at the data or the given elements and see how these may be determined by the other parties to the transactions we study. This means that we try to change the *exogenous* variables, i.e., those given from the outside, into *endogenous* variables, i.e., variables which are determined within the system. If we do this for more and more variables, we will get more general, or more *closed*, models. These models can be taken as theories about how a system develops through the interaction of the decisions made by different people. This means that by broadening and generalizing the basis of the decision problems we get an economic theory.

This approach may be the natural one for students of business administration. To economists it may be more natural to take the classical economic theory as the starting point. In this theory there is no room for uncertainty. The theory assumes that people decide how to consume, produce, and invest with full knowledge of what the outcome of their decisions will be. Uncertainty is either ignored or explicitly "assumed away." It is obvious that the resulting theory is not very realistic, and probably not very useful. In order to get a realistic economic theory, we must bring in uncertainty, probably as an essential element of the theory.

1.15. Modern economic theory and decision theory both have the reputation of being very "mathematical." There are some good reasons for this, but the essential ideas in these theories can be mastered without much technical knowledge of advanced mathematics. What is required to understand and use these theories may be what Luce and Raiffa ([2] Preface) call "that ill-defined quality: mathematical sophistication," but it may also be just a good deal of clear logical thinking. This is a prerequisite of a good mathematician, but mathematicians are not the only people who can think straight—although they may often claim so.

Much of the mathematics we find in the literature on decision theory is of computational value, i.e., algorithms for solving special classes of problems. This is, of course, important, particularly if we want to apply the theory in practice, but it is a purely technical job. If we have a good general understanding of the problem and its whole background, we can nearly always find a solution—by consulting an expert or searching reference books.

1.16. In the following chapters we shall argue that the really important thing is to *formulate* the problem so that it can be solved mathematically. In order to do this intelligently, we must have some general knowledge of what mathematics is, and what it can do, but it is not necessary to have all the details at our fingertips.

The real problems—in business and industry—are usually very complicated, so that we must reformulate them, or simplify them, before we can solve them. The real art in this field is to reduce the problem to something we can manage, without losing anything essential to the problem we originally set out to solve.

It has been said that much work in Operation Research "gives the right answer to the wrong question." The mathematics in such work is usually perfect, but the problem may have been simplified to such an extent that it bears no relation to the real-life problem we really wanted to solve. The world seems to be full of enthusiasts who love mathematical manipulations and who want to compute at any cost, and as soon as possible. Very often it would be useful if these people would cool down a little and ask *what* they want to compute, instead of the inevitable *how*.

1.17. Most of the ideas we shall discuss in this book have their origin in *game theory*, created by Von Neumann and Morgenstern. It may, therefore, be appropriate to close the chapter by quoting one of their remarks about the uncritical use of mathematics in economics and other social sciences: "The underlying vagueness and ignorance has not been dispelled by the inadequate and inappropriate use of a powerful instrument that is very difficult to handle" ([3] pp. 4–5). In this book we shall concentrate on the basic "vagueness and ignorance" and seek to avoid unnecessary exhibition of the "powerful instrument" mathematics really is.

REFERENCES

[1] Howard, R. A.: *Dynamic Programming and Markov Processes*, The MIT Press, 1960.
[2] Luce, R. D. and H. Raiffa: *Games and Decisions*, Wiley, 1957.
[3] Neumann, J. von and O. Morgenstern: *Theory of Games and Economic Behavior*, 2nd ed., Princeton University Press, 1947.

Chapter II

Economic Decisions under Uncertainty

2.1. When there is no uncertainty, the decision problem is in some ways trivial. Let us assume that we have to select one of the two actions A and B, either one costing us no effort.

If we decide on A we get $100.
If we decide on B we get $200.

In this situation we obviously select B. This seems trivial because in economics we take it for granted that people prefer $200 to $100.

Practical decision problems under full certainty are not so simple, but in principle they can be reduced to this pattern, and then the decision is trivial. This reduction may, however, be very complicated. A and B may, for instance, be two production schedules, i.e., two ways of producing a certain amount of goods. It may be trivial to say that we want to produce these goods at the least possible cost, but it may be very difficult to determine the schedule which actually gives the minimum cost.

2.2. In the following we shall not discuss these difficulties. We want to discuss the particular problems connected with uncertainty, and in order to do this, we have to clear away difficulties which are irrelevant to our purpose. In doing so, we dismiss a whole range of problems which have played an important part in *management science*, in fact, problems which constitute the very core of this subject.

It is important that we should realize what we are doing, and we should also realize that this kind of abstraction or "assuming away" is necessary in order to come to grips with the deeper problems. It will not make much sense to study decisions under uncertainty unless we assume that we know how to make decisions under full certainty. This may be an unrealistic assumption, but we must accept it—at least temporarily—in order to make any headway with the problems which really interest us.

2.3. Before we attack the problem of uncertainty, it may be useful to consider a classical problem, which we shall return to several times in later chapters. In economic theory, it is usual to assume that there exists a *production function*

$$y = f(x_1, \ldots, x_n).$$

The interpretation is that if we use the quantities $x_1, \ldots, x_n$ of the different inputs (factors of production), we obtain an output of y units of the finished product. This model is in itself an abstraction, or an idealized picture of the real world.

The producer, or the owner of the factory with this production function, has to decide on an "input vector," or an n-tuple $(x_1, \ldots, x_n)$. If the price of output is p_0 and the prices of inputs are $p_1, \ldots, p_n$, the profit of the producer will be

$$P = p_0 f(x_1, \ldots, x_n) - \sum_{i=1}^{n} p_i x_i.$$

Classical economic theory took it as obvious, or trivial, that the producer wanted to maximize his profits. Hence, the "management problem" is to determine the input vector $(x_1, \ldots, x_n)$ which maximizes P.

This problem becomes meaningless once we bring uncertainty into the model. The most obvious way of doing this may be to assume that there is some uncertainty about the prices $p_0, p_1, \ldots, p_n$. We may, however, come closer to the real problem if we assume that the relation between inputs $x_1, \ldots, x_n$ and output y is of a stochastic nature. Every farmer will know that the production function really is a stochastic relation, and it is likely that most engineers will agree.

2.4. Let us now return to the problems of uncertainty. We saw in 2.1 that we obviously prefer action B to action A if B gives us $200 and A $100 with certainty.

The situation is not so obvious if A gives us $100 with certainty, and if B gives us either $500 if a tossed coin falls heads or nothing if the coin falls tails.

The only thing which is obvious is that in this situation we are confronted with quite a new problem. $100 may be good, but is this better than a 50–50 chance of getting $500? There is no obvious way in which we can decide. It seems that we must "decide for ourselves," as we cannot solve the problem by appealing to general rules or principles.

2.5. To bring out the essentials in this new problem, let us consider a slightly more general version of the model. Let us assume that we have to decide on one of the actions $A_0, A_1, \ldots, A_n, \ldots$. If we decide on A_n, we will get

either $$S_n$ with probability P_n
or nothing with probability $1 - P_n$.

Let us assume that the two functions S_n and P_n are given by Table 1. The table is based on the formulae

$$P_n = \frac{10}{10 + n} \quad \text{and} \quad S_n = 10\left(\frac{n}{10}\right)^{0.9}$$

so we can interpolate and extrapolate if we think that this will make the decision easier.

TABLE 1

	P_n	S_n
A_0	1	0
A_1	0.9	2
A_{10}	0.5	10
A_{20}	0.33	19
A_{30}	0.25	27
A_{40}	0.20	35
A_{90}	0.10	72
A_{190}	0.05	140
A_{990}	0.01	625
A_{9990}	0.001	5,000
A_{99990}	0.0001	40,000
A_{999990}	0.00001	250,000
$\vdots$	$\vdots$	$\vdots$

2.6. In this table, the only obvious thing is that we should never decide on A_0. With this decision we are certain of getting nothing. All the other decisions give us a chance of getting something. The question is really how far down in the table we should go.

A_1 gives us a very good chance of getting $2, which is much better than the certainty of getting nothing.

A_{10} gives us a 50–50 chance of getting $10, an amount which will pay for a good dinner or for a textbook which may explain how one should deal with such decision problems.

A_{999990} offers a chance of getting really rich, although with very long odds.

2.7. It is not likely that in real life we will be confronted with a decision problem of this kind. It is, however, worthwhile making a mental effort to try to take this problem seriously and to figure out how we actually would decide in a situation like this. Unless we can handle such simple problems, it is preposterous to claim that we can make intelligent decisions in far more complicated situations. So let us take a good look at the table and ask the following questions: What kind of criterion would we use? How would we decide?

The classical rule is that we should seek to maximize the *expected gain*, or the "mathematical expectation." In our case this means that we should pick the action A_n, for which the product $P_n S_n$ takes its greatest value. We have

$$E_n = P_n S_n = \frac{100}{10 + n} \left(\frac{n}{10}\right)^{0.9} = \frac{10^{1.1} n^{0.9}}{10 + n}.$$

Taking E_n as a continuous function of n, we differentiate and find

$$\frac{dE_n}{dn} = \frac{10^{1.1}}{n^{0.1}(10 + n)^2} \{9 - 0.1n\}$$

From this expression it is easy to see that E_n takes its maximum value for $n = 90$. This means that our best choice should be A_{90}, i.e., to take 1 chance in 10 of gaining \$72. From inspection of the table it is not clear why this should be the best action, and the classical rule appears completely arbitrary. If this decision has to be made only once, we can think of a number of reasons for selecting one of the other actions.

2.8. The real justification for using the expected gain as a "decision criterion" is, of course, the *Law of Large Numbers*. If we have to pick an action in our table 10,000 times, for instance, every day in 30 years, it is "practically certain" that consistent choice of A_{90} will give us an average gain of $P_{90}S_{90} = 7.2$, and hence the greatest total gain. This will hold also under more general conditions. If we have to make a large number of such choices, not necessarily from the same table, we will do best in the "long run" if we always select the action which gives the greatest expected gain.

However, whenever we can appeal to the Law of Large Numbers, we are justified in using the term "practically certain." This means, of course, that the uncertainty has somehow disappeared and that we have been led back to the relatively trivial problem discussed in 2.1. In this book we want to study decision problems and situations where it is not possible to "assume away" uncertainty in this manner without losing some essential aspect of the original problem. This means that we will have to focus our attention on situations where the Law of Large Numbers does not apply, i.e., situations where it is not sufficient to consider only the long-run outcome of our decisions. It is likely that such situations are of importance in practice, since, as more than one economist has remarked, "In the long run we will all be dead."

2.9. The idea that intelligent (or rational) people ought to make their decisions so that the mathematical expectation of the gain is maximized goes back to the beginning of probability theory. This is quite natural because the original purpose of probability theory was to determine how a gambler should play in order to do best in *the long run*. If we spend the whole night playing dice, the Law of Large Numbers may be useful.

Daniel Bernoulli pointed out as early as 1732 that maximizing expected gain is not a golden rule of universal validity. His famous counterexample is the so-called *St. Petersburg Paradox* [1], which is illustrated by the following game:

A coin is tossed until it falls heads. If heads occurs for the first time at the nth toss, the player gets a prize of 2^n dollars (ducats in Bernoulli's description of the game) and the game is over. The probability that the coin falls heads for the first time at the nth toss is obviously $(\frac{1}{2})^n$. Hence the expected gain in this game is

$$\sum_{n=1}^{\infty} 2^n(\tfrac{1}{2})^n = \infty,$$

since it is theoretically possible that the game will go on for ever.

A "game" which will give $1 million, payable with certainty, has an expected gain of only $1 million. This means that a person who follows the rule of selecting the game with the greatest expected gain will take an opportunity of playing a St. Petersburg game rather than a million dollars. This is obviously nonsense, since we know from experience that people don't make their choices in this way.

2.10. Let us now return to Table 1. It is clear that the A's in the table can be interpreted as tickets in different lotteries. Our problem is then to select the lottery in which to take a ticket. In our example there is just one prize in each lottery. It is desirable to consider a more general situation.

Let us therefore consider the following model. Each A_n corresponds to a lottery, where the prizes are

$$S_n{}^0, S_n{}^1, S_n{}^2, \ldots.$$

These can be won with probabilities

$$P_n{}^0, P_n{}^1, P_n{}^2, \ldots,$$

where $\sum_i P_n{}^i = 1$ for all n.

In the following we shall find it convenient to use the notation

$$x_i = S_n{}^i$$
$$f_n(x_i) = P_n{}^i.$$

We can then consider $f_n(x)$ as a discrete probability distribution; i.e., $f_n(x_i) = $ the probability of winning a prize x_i in lottery n.

Our problem can be formulated as follows: We consider a set of lotteries. Each element A_n of the set is represented by a probability distribution $f_n(x)$. Our problem is then to select the best element in this set, i.e., the lottery, in which we want to take a ticket.

In the example in 2.5 each lottery has only two prizes, so that it must be represented by a probability distribution which is different from zero only for two values; i.e., we have

$$f_n(S_n) = P_n$$
$$f_n(0) = 1 - P_n$$

2.11. We can now make the following observations:

(i) The decision is not trivial. If we have the choice of either a 10-dollar bill or a 20-dollar bill, it is obvious that we take the 20-dollar bill. In the more general case, it is not obvious which lottery ticket we should take.

(ii) Any decision problem under uncertainty can in principle be reduced to this pattern, i.e., to choose the best from a set of available probability distributions. This reduction may be a formidable task, and it may be next to impossible to carry it out in practice. The exact form of the distribution may not be known, but for the time being we shall ignore this difficulty.

To give some concrete contents to these observations, let us consider a businessman who has to choose either action *A* or action *B* in a complicated situation with uncertainty of different kinds. If he engages a management consultant to study the problem, the consultant can possibly carry out the technical job of reducing the problem to the pattern we have considered. The result may be as follows:

Action *A*: A 50–50 chance of either a profit of $10,000 or a profit of $12,000.

Action *B*: A 50–50 chance of either a profit of $40,000 or a loss of $2,000.

A more realistic result may be that action *A* will lead to a profit distribution as indicated in Fig. 1, and action *B* will give a distribution of the shape indicated in Fig. 2.

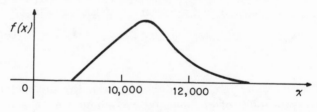

Figure 1

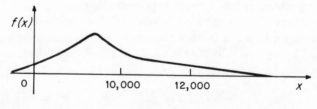

Figure 2

2.12. It is important to realize that in many situations an honest consultant *cannot go beyond presenting the result in this form*. He may add that the businessman must decide for himself how he feels about taking a risk and that the ultimate decision is his and not that of the consultant.

The businessman will usually want something more from an expensive consultant. He will want a clear recommendation as to whether he should take action *A* or action *B*. If the consultant gives an unambiguous recommendation, he must either know how the businessman feels about risk, or he must base his recommendation on some general principle, which may be of moral or ethical nature. He may, for instance, recommend action *A* because he thinks it is wrong to gamble, even against favourable odds, or because he thinks that his client is in a situation where he "should not" take risks such as those associated with action *B*. If the consultant is pressed to explain his recommendation of action *A*, he will probably fall back and state that, in his experience, people of "sound judgement" prefer action *A* to action *B* in situations similar to that of his client. This means that the recommendation, which is inherently normative, is reduced to a descriptive statement about the behavior of a certain group of people. It is still up to our businessman to decide for himself whether he should follow the example of these people or not.

There is much confused thinking on these matters in business, and this confusion sometimes finds its way into technical literature. The reason is probably that businessmen—naturally enough—prefer a recipe for success to a description of their problem.

2.13. We can find good examples of the problems we have discussed in the insurance world. As a realistic example, let us consider a big ship—an ocean liner—insured for $10 million against an annual premium of $300,000. If things go well, this insurance contract will give the insurance company a clear profit of $300,000 each year, and that means something, even to a big company. On the other hand, if the ship gets lost at sea, the company will have to pay $10 million, and that may create serious difficulties which may take the company years to overcome.

Normally, an insurance company will not hold a contract of this kind on its own account. Usually a substantial part of the contract will be *reinsured*, i.e., passed on to another insurance company. If our company *retains* only 10%, i.e., reinsures 90%, the risk situation will be changed to a loss of $1 million or a profit of $30,000 (provided that reinsurance is obtained on "original terms"). The problem is then to decide how much should be reinsured; this is a situation similar to the one illustrated by Table 1.

2.14. The President or the Board of our insurance company may well ask their actuary or a statistical consultant how they *should* reinsure the contract. The actuary in all honesty should reply that the problem is reduced to its simplest possible form and that the Board must decide for itself when it has made up its mind how it feels about running risks.

Usually, boards don't accept such answers from highly paid actuaries. They expect the actuary to give his answer only after elaborate computations. If he is a good actuary, he should process all relevant statistics of shipwrecks and consult with engineers about the technical properties of the ship.

However, what can be the outcome of such computations? The result can be little more than a statement to the effect that the best estimate of the probability that the ship will be lost is 0.011. To this one can add some qualification, for instance that the confidence interval of the estimate is 0.015–0.009 at the 1% significance level, and 0.012–0.010 at the 5% level. No amount of computation can give more information.

We should, however, ask ourselves if such information is useful, that is, if it really will help if these simple results are imbedded in a 50-page technical report. In one respect the situation is quite clear. What the Board really wants to know is *whether the ship will sink or not*. Nobody but an astrologer can tell them that.

2.15. It is told that a well-known professor of insurance once replied in this way when he was consulted by an insurance company. The company considered this for some weeks, and then invited him to join the Board. The really amusing part of the story is that the company thought that this was just what the professor wanted in the first place, and that he gave an "evasive" answer in order to get a seat on the Board. They were convinced that there was a unique correct solution to their problem—presumably known to a professor.

We can learn a lot about business decisions from this story—even if it is not true. The story illustrates the shrewd thinking and suspicion of ulterior motives which must be very useful, if not essential, to success in business. The story, however, also gives an example of the rather naive wishful thinking which we often find in business circles—and in business literature.

A businessman may often have to make difficult decisions in complicated situations. In such cases he can call upon technicians or "experts" who may be able to clarify the situation, but they cannot, in general, tell the businessman that a particular decision is the only right one. The ultimate decision as to what risks should be taken must remain the responsibility of the businessman himself. Businessmen often seem to find it difficult to accept this fact of life. A psychologist may interpret this as a search for a father figure in the person of the expert.

2.16. Let us now return to the insurance example given in 2.13. In that example we took the premium of $300,000 as given. This means that we ignored the decision problem which in practice is considered as most important by insurance companies, namely what premium to *quote* for a proposed insurance coverage.

By statistical analysis and computations it may be possible to sort out the problem so that it seems to fit into the simple pattern we have discussed. However, if we force the problem into this pattern, we may lose something essential. It seems a little naive to quote a premium just on the merits of the contract, without considering both how much the ship-owner is willing to pay for insurance coverage and the premium which competing companies may quote. If the ship-owner is prepared to pay $400,000 for insurance, and no other company is willing to take the risk at this price, there is no reason why our company should quote a lower premium.

This illustrates the point made in Chapter I, that we cannot—or should not—consider a decision problem in isolation. We must consider all potential parties to the transactions involved, and this means that we have to see the particular problem in the framework of a more general theory, which it is natural to call the "Economic Theory of Uncertainty."

We arrive at a similar conclusion if we consider reinsurance. In our example we assumed that there would always be other companies willing to take over the business we did not want on "original terms," i.e., on the terms on which we originally gave coverage to the ship-owner. It is not obvious that this assumption holds, and it may be advisable to investigate it carefully before we start computing in great detail how much business our company should give away to these other companies. It may turn out that they are not interested in taking over the business on these terms.

2.17. The decision problem under full certainty is not always as trivial as when we have to choose either a 10-dollar bill or a 20-dollar bill. To illustrate the point, let us assume that we have to select one of two "market baskets":

Basket 1, $\{g, v\}$,   contains one bottle of gin and one bottle of vermouth
Basket 2, $\{2g, 2v\}$, contains two bottles of gin and two bottles of vermouth.

This decision is trivial to most people, since we are discussing free gifts. To establish this formally, we need an assumption or an axiom to the effect that "more is better" of any good. This tautology is not trivial. It implies that there is no saturation of needs.

The decision is not trivial if we have to pick one of the baskets

$$\{4g, v\}   \text{or}   \{3g, 2v\}.$$

It is intuitively clear that the decision must depend on "subjective elements," such as personal taste, or, to be specific, on how we like to mix a dry martini.

Decisions of this kind are clearly very important in practice. In fact, economic activity consists in some sense in selecting the best from a set of generalized "market baskets."

2.18. To construct an economic theory it is essential to solve the problem we have outlined. Since no obvious solution offers itself, economists have simply assumed that the problem can be solved—an assumption necessary in order to make progress. They have assumed that if a person—a consumer—has the choice of one of two baskets

$$\{x_1g, y_1v\} \quad \text{or} \quad \{x_2g, y_2v\}$$

he has some rule which enables him to decide which basket is best, or which basket he prefers.

Usually one makes a stronger assumption: one assumes that a person, when confronted with a whole range of baskets, can arrange them *in order of preference*, i.e., that he can indicate not only his first choice, but also the second, third, etc. This assumption can also be proved as a theorem from a more basic assumption of *transitivity of preferences*, that is, if

A is preferred to B and B is preferred to C,

then

A is preferred to C.

Mathematically this means that we assume a complete ordering over the set of ordered pairs (x, y), each representing a basket of bottles.

In economic textbooks this preference ordering is usually illustrated by an *indifference map*, which by some intuition we can generalize to more than two dimensions, i.e., to baskets with more than two different goods.

2.19. As long as we consider only baskets which contain a finite number of unbroken bottles, it is obvious that the preference ordering can be represented by a utility index, or a *utility function* $u(x, y)$, such that

$$u(x_1, y_1) > u(x_2, y_2)$$

if and only if basket $\{x_1g, y_1v\}$ is preferred to basket $\{x_2g, y_2v\}$. This will hold more generally if we compare baskets with an arbitrary, finite number of different commodities. In this case the number of possible baskets will be finite, say N. We can then represent the preference ordering by a utility index by assigning the utility N to the best basket, $N - 1$ to the second best, and so on. This representation is obviously not unique, since we could have assigned the utilities $N^2, (N - 1)^2, \ldots$.

If we bring in infinity in some way, for instance by assuming that the goods in the basket are "infinitely divisible," it may not be possible to represent a preference ordering by a utility function. In practical work we can, of course, avoid infinity altogether. This will, however, have some undesirable consequences; for instance, we may not be able to use the differential calculus so dear to economists. In order to bring uncertainty into the model, we must, however, be able to handle infinity, as we shall see in the following chapters.

2.20. The problem of representing a general preference ordering by a utility function was first discussed by Wold [3]. A particularly elegant formulation of the conditions under which such a representation is possible is given by Debreu [2].

The simplest example of a preference ordering which cannot be represented by a utility function is the *lexicographic* ordering. To illustrate this, let us consider a "market basket" containing x dollar bills and y tickets in the Irish Sweepstake. It is possible to rank such baskets as follows:

(x_1, y_1) is preferred to (x_2, y_2) if and only if $x_1 > x_2$, i.e., regardless of the values of y_1 and y_2.
(x_1, y_1) is preferred to (x_1, y_2) if and only if $y_1 > y_2$.

This means that our person will always prefer the basket with more dollars, regardless of how many tickets it contains. Only if the number of dollars is the same in two baskets, will he consider the number of sweepstake tickets. It is clear that this well-defined preference ordering cannot be represented by a utility function $u(x, y)$. Our first condition implies that the function is independent of y, but the second condition assumes that such a dependence exists.

In elementary economic analysis one should probably not attach much importance to such examples. In most cases it seems fairly safe to assume that preference orderings can be represented by a utility function in several different ways. There may exist persons with more or less "perverse" preference orderings who do not fit our model, but we can ignore them as long as we do not set our level of aspiration too high.

2.21. A basic assumption in the following chapter is that some kind of "trade-off" will always be possible. Formally we can express this by assuming the so-called *Axiom of Archimedes*. This means in our example that if we have

$$(x_1, y_1) \text{ preferred to } (x_2, y_2)$$

we can always reverse the preference by increasing y_2; i.e., there exists a $y > y_2$ such that

$$(x_2, y) \text{ if preferred to } (x_1, y_1).$$

This means that a loss of some units of one commodity can always be compensated by a gain of some units of another commodity or, to put it another way, *everything has its price*. It may be tempting to define economics as the science of things which have a price, in a very general sense. Questions of life and death and ethical principles like an absolute aversion to gambling would then be considered as belonging to the more general social sciences—outside the narrow subject of economics proper.

REFERENCES

[1] Bernoulli, D.: "Specimen Theoriae Novae de Mensura Sortis," St. Petersburg 1738. English translation: *Econometrica*, 1954, pp. 23–36.

[2] Debreu, G.: "Representation of a Preference Ordering by a Numerical Function," pp. 159–166 in *Decision Processes*, edited by Thrall, Coombs, and David. Wiley, 1954.

[3] Wold, H.: "A Synthesis of Pure Demand Analysis," *Skandinavisk Aktuarietidskrift*, 1943, pp. 85–118, pp. 220–263, and 1944, pp. 69–120.

Chapter III

The Bernoulli Principle

3.1. In 2.10 we saw that a decision under uncertainty consisted in selecting a probability distribution from a given set of such distributions. A rational decision-maker will, by definition, select the "best" of the available distributions. This means that a rational theory for decisions under uncertainty must be based on preference orderings over sets of probability distributions, i.e., over sets where the elements $f_1(x), \ldots, f_n(x), \ldots$ are probability distributions.

As a concrete interpretation, we can think of these elements as tickets in different lotteries. We can also think of them as different investments, defined so that investment n will give as gain a stochastic variable x with probability distributions $f_n(x)$.

To simplify matters, we shall for the time being consider only discrete distributions. We can then interpret $f_n(x_1), f_n(x_2), \ldots, f_n(x_i), \ldots$ as the probabilities that investment n will give the gains $x_1, x_2, \ldots, x_i, \ldots$ respectively. We must obviously have

$$\sum_i f_n(x_i) = 1 \quad \text{for all } n.$$

Marschak [8] has introduced the convenient word *prospect* for the elements of the set which we want to consider, and we shall use this word in what follows.

3.2. A discrete prospect can be completely described by a sequence

$$\ldots, f(-2), f(-1), f(0), f(1), f(2), \ldots, f(x), \ldots$$

where $f(x)$ is the probability that the gain will be x dollars (or x cents or, if necessary, fractions of a cent).

We seek a preference ordering over the set of prospects, and we shall assume that this ordering can be represented by a utility function—or in this more general case, by a utility *functional*. This means that we want to associate a real number $U\{f\}$ with each probability distribution $f(x)$ such that

$$U\{f_i\} > U\{f_j\}$$

if and only if the prospect $f_i(x)$ is preferred to $f_j(x)$. In mathematical terms our problem is to find a *mapping* from the space of all discrete probability distributions to the real line.

Formally we can consider the functional $U\{f\}$ as an ordinary utility function of the type we studied in 2.19. We can then write

$$U\{f\} = u[\ldots, f(-1), f(0), f(1), \ldots, f(x), \ldots].$$

The only difficulty is that u in general will be a function of an infinite number of variables. Modern mathematics is quite capable of dealing with such functions, but only by using tools which so far have found few applications in economics.

3.3. Before we attack our problem in earnest, we shall make some general remarks. First we shall note that the familiar theoretical framework from classical economics can be applied to the economics of uncertainty, provided that we can pass from the finite to the infinite. Second we shall note that there is a very peculiar algebra involved in this approach to the economics of uncertainty.

If we add two market baskets, $\{x_1 g, y_1 v\}$ and $\{x_2 g, y_2 v\}$, of the kind studied in Chapter II, we obtain another basket $\{(x_1 + x_2)g, (y_1 + y_2)v\}$, which contains just the two commodities found in the original baskets.

If we add two stochastically independent prospects, described by

$$f_1(0) = \tfrac{1}{2}, f_1(1) = \tfrac{1}{2} \quad \text{and} \quad f_2(0) = \tfrac{1}{2}, f_2(1) = \tfrac{1}{2},$$

we obtain a three-point probability distribution described by

$$f_3(0) = f_1(0)f_2(0) = \tfrac{1}{4},$$
$$f_3(1) = f_1(0)f_2(1) + f_1(1)f_2(0) = \tfrac{1}{2},$$
$$f_3(2) = f_1(1)f_2(1) = \tfrac{1}{4}.$$

In vector notation this can be written

$$\{\tfrac{1}{2}, \tfrac{1}{2}, 0, \ldots\} + \{\tfrac{1}{2}, \tfrac{1}{2}, 0, 0, \ldots\} = \{\tfrac{1}{4}, \tfrac{1}{2}, \tfrac{1}{4}, 0, \ldots\}.$$

In terms of the classical economic theory we can then interpret the original prospects as market baskets containing the two "commodities": (1) a chance of gaining nothing and (2) a chance of gaining 1. Both baskets contain the "quantity" $\tfrac{1}{2}$ of these two commodities. When we add the two baskets, we obtain a new commodity, a chance of gaining 2, in a quantity $\tfrac{1}{4}$.

Mathematically this means that addition of market baskets in classical economics corresponds to the operation of *convolution* of probability distributions in the economics of uncertainty. This is, however, valid only under the assumption that the two prospects are stochastically independent. In general, addition of prospects has no meaning unless we specify the stochastic dependence between them. This gives us some idea of the nature of the difficulties which we will encounter in the economics of uncertainty.

Third we note that some decisions under uncertainty are obvious, but others must depend on some "subjective elements." To illustrate this, let us consider the three prospects

    f_1: Gain 0 or \$1, each with probability $\frac{1}{2}$, or in vector notation $\{\frac{1}{2}, \frac{1}{2}, 0, 0, \ldots\}$

    f_2: Gain 0 or \$2, each with probability $\frac{1}{2}$, or $\{\frac{1}{2}, 0, \frac{1}{2}, 0, \ldots\}$

    f_3: Gain 0 with probability $\frac{1}{4}$, or gain \$1 with probability $\frac{3}{4}$. In vector notation $\{\frac{1}{4}, \frac{3}{4}, 0, 0, \ldots\}$.

Here it is clear that both f_2 and f_3 should be preferred to f_1, just as we found in 2.17 that $\{2g, 2v\}$ was preferred to $\{g, v\}$. It is, however, not clear whether f_2 should be preferred to f_3 or not.

If a person insists that he prefers f_3 to f_2, we have to accept this as an expression of his personal *attitude to risk*. This corresponds to the statement $\{4g, v\}$ preferred to $\{3g, 2v\}$, which in 2.17 we accepted as an expression of personal taste.

3.4. For simplicity we shall in the following consider only prospects where all gains are non-negative and finite.

Let us first consider the set of all prospects $f(x)$, defined by discrete probability distributions over the domain $0 \le x \le M$; where x and M are integers. We want to establish a preference ordering over this set, and we shall begin by considering some general conditions which the ordering method or "selection rule" must satisfy if it is to be acceptable to a rational person.

The first of these conditions we formulate as an axiom.

Axiom 1: To any prospect $f(x)$ in the set there corresponds a certainty equivalent $\bar{x}$.

This is really the Axiom of Archimedes, which we mentioned in 2.21. In popular terms $\bar{x}$ is the lowest price at which we will sell the prospect, or the highest price we are willing to pay for it. In more precise terms, it is the same to me (I am indifferent) whether I own the prospect $f(x)$ or an amount of cash equal to $\bar{x}$. We shall write

$$(1, \bar{x}) \sim f(x)$$

for the relation which defines the certainty equivalent.

3.5. The set we consider includes all prospects of the binary type in which the only two possible outcomes are

M with probability p.
0 with probability $1 - p$.

In the following we shall let (p, M) stand for such a binary prospect. It follows from our Axiom 1 that these prospects all have their certainty equivalents; i.e., for any p, there exists a number x_p, so that

$$(1, x_p) \sim (p, M).$$

We then introduce a second axiom:

Axiom 2: As p increases from 0 to 1, x_p will increase from 0 to M.

This means that for all integral values of x: $0, 1, \ldots, r, \ldots, M$ there exist values of p: $p_0, \ldots, p_n$ such that

$$(1, r) \sim (p_r, M)$$

and $p_r > p_s$ if and only if $r > s$.

3.6. $f(x)$ is an abbreviated way of describing a prospect. To give a full description, we would have to state that the prizes $0, 1, \ldots, r, \ldots, M$ can be won with the probabilities $f(0), \ldots, f(r), \ldots, f(M)$ respectively.

Let us now replace the prize r with the equivalent binary prospect (p_r, M). This will give us a modified prospect $f^{(r)}(x)$, where there is no prize equal to r. It is easy to see that we have

$$f^{(r)}(0) = f(0) + f(r)(1 - p_r)$$
$$f^{(r)}(1) = f(1)$$
$$f^{(r)}(2) = f(2)$$
$$\cdot \quad \cdot \quad \cdot \quad \cdot \quad \cdot \quad \cdot$$
$$f^{(r)}(r) = 0$$
$$\cdot \quad \cdot \quad \cdot \quad \cdot \quad \cdot \quad \cdot$$
$$f^{(r)}(M) = f(M) + f(r)p_r.$$

Since $(1, r) \sim (p_r, M)$ it is natural to assume that a rational person is indifferent as to whether his prospect is modified in this way or not. Formally we express this as an axiom:

Axiom 3: $f(x)$ and $f^{(r)}(x)$ as defined above have the same certainty equivalent.

3.7. Let us now apply this axiom to all the prizes except 0 and M. We then get a prospect of the type (P, M), which has the same certainty equivalent as the original prospect $f(x)$. P is determined by

$$P = p_1 f(1) + p_2 f(2) + \cdots + p_{M-1} f(M - 1) + f(M).$$

Since $p_0 = 0$ and $p_M = 1$ by Axiom 2, we can write

$$P = \sum_{x=0}^{M} p_x f(x).$$

From this we can obtain a complete preference ordering over our set of prospects. For two arbitrary prospects $f(x)$ and $g(x)$ we can compute the corresponding prospects (P_f, M) and (P_g, M) and their certainty equivalents. The ordering is then that $f(x)$ is preferred to $g(x)$ if and only if $P_f > P_g$ (or equivalently if $f(x)$ has the greater certainty equivalent).

3.8. If we want to represent this preference ordering by a utility functional, we can define

$$U\{f(x)\} = P_f = \sum_{x=0}^{M} p_x f(x).$$

If we write $p_x = u(x)$, our formula can be written

$$U\{f(x)\} = \sum_{x=0}^{M} u(x)f(x).$$

Our set of prospects includes degenerate probability distributions representing the cases where the gain can take only one value with probability one, i.e., distributions of the type

$$e_r(x) = 0 \qquad x \neq r$$
$$e_r(x) = 1 \qquad x = r.$$

If we apply our formula to this prospect, we obtain

$$U\{e_r(x)\} = u(r).$$

This means that $u(x)$ is the utility assigned to the prospect which will give us the amount x of money with certainty. From this it follows that the function $u(x)$ can be interpreted as the *utility of money*—a concept which played a very important part in classical economic theory.

Some people consider this very surprising and have created considerable confusion trying either to explain or to deny the validity of the result.

3.9. To explain the result, let us retrace our steps:

We observed that people seem to be able to decide on a choice when they have to select an element from a set of prospects.

We next assumed that people who make such choices must have some rule which enables them to decide when one prospect is better than another. We started with a completely open mind as to the nature of this selection rule.

We then formulated three simple conditions which the rule ought to satisfy if it was to be acceptable to intelligent logical people.

We then found that a selection rule which satisfies these conditions can be *represented* by a function $u(x)$.

The rule as to when prospect $f(x)$ is preferred to $g(x)$ can be given in different ways. If the rule satisfies our three axioms, there will always exist a function $u(x)$, so that application of the rule consists in computing the two sums

$$\sum u(x)f(x) \quad \text{and} \quad \sum u(x)g(x).$$

The prospect which gives the largest sum is then the "best," or most preferred.

This is *not* the only way in which we can describe a selection rule which satisfies the three axioms. In most applications this is, however, by far the most convenient description.

3.10. If a preference ordering can be represented by a function $u(x)$, it can also be represented by the function

$$v(x) = au(x) + b,$$

where a and b are constants and $a > 0$, since obviously

$$\sum u(x)f(x) > \sum u(x)g(x)$$

implies

$$\sum v(x)f(x) > \sum v(x)g(x).$$

We shall next prove that if two utility functions represent the same preference ordering they must satisfy an equation of this form.

If the two utility functions $u(x)$ and $v(x)$ represent the same preference ordering, the inequalities above must both be satisfied, or both reversed, for any pair of probability distributions $f(x)$ and $g(x)$. This means that the two sums

$$\sum u(x)\{f(x) - g(x)\}$$

and

$$\sum v(x)\{f(x) - g(x)\}$$

must have the same sign, or that the product of the two sums must be non-negative.

Let us now for the sake of simplicity consider only a three-point distribution, so that the first sum can be written

$$u(x_1)\{f(x_1) - g(x_1)\} + u(x_2)\{f(x_2) - g(x_2)\} + u(x_3)\{f(x_3) - g(x_3)\}$$

or with a simpler notation

$$u_1(f_1 - g_1) + u_2(f_2 - g_2) + u_3(f_3 - g_3).$$

As $f_3 = 1 - f_1 - f_2$ and $g_3 = 1 - g_1 - g_2$ this expression becomes

$$(u_1 - u_3)(f_1 - g_1) + (u_2 - u_3)(f_2 - g_2).$$

Similarly we find for the second sum

$$(v_1 - v_3)(f_1 - g_1) + (v_2 - v_3)(f_2 - g_2).$$

Principle

Hence the two utility functions u and v can represent the same preference ordering only if

$$\{(u_1 - u_3)(f_1 - g_1) + (u_2 - u_3)(f_2 - g_2)\}$$
$$\times \{(v_1 - v_3)(f_1 - g_1) + (v_2 - v_3)(f_2 - g_2)\} \geq 0$$

or

$$(u_1 - u_3)(v_1 - v_3)(f_1 - g_1)^2 + [(u_1 - u_3)(v_2 - v_3) + (u_2 - u_3)(v_1 - v_3)]$$
$$\times (f_1 - g_1)(f_2 - g_2) + (u_2 - u_3)(v_2 - v_3)(f_2 - g_2)^2 \geq 0$$

for all values of $(f_1 - g_1)$ and $(f_2 - g_2)$. This condition can be satisfied only if the left-hand side is a complete square, i.e., if

$$(u_1 - u_3)(v_2 - v_3) = (u_2 - u_3)(v_1 - v_3)$$

or

$$\frac{v_2 - v_3}{v_1 - v_3} = \frac{u_2 - u_3}{u_1 - u_3}.$$

From this expression we find

$$v_2 = \frac{v_1 - v_3}{u_1 - u_3} u_2 + \frac{u_1 v_3 - u_3 v_1}{u_1 - u_3}.$$

Let us now reintroduce our original notation. If we consider x_1 and x_3 as fixed, and write x for x_2, we obtain

$$v(x) = \frac{v(x_1) - v(x_3)}{u(x_1) - u(x_3)} u(x) + \frac{u(x_1)v(x_3) - u(x_3)v(x_1)}{u(x_1) - u(x_3)}.$$

This is clearly a relation of the form $v(x) = au(x) + b$, which the two utility functions must satisfy if they are to represent the same preference ordering.

It is not very difficult to show that the same condition must hold also for prospects with more than three possible outcomes. We can do this by using the general theory of positive-definite quadratic forms, or we can accept the intuitive argument that an increase in the number of outcomes cannot weaken the restrictions imposed on utility functions which shall represent the same preference ordering.

We can express our results as a theorem:

Theorem: A preference ordering which satisfies Axioms 1, 2, and 3 can be represented by a utility function, unique up to a positive linear transformation.

3.11. The method we used to derive the theorem from the axioms is valid only for prospects with a finite number of different prizes. If we study prospects with an infinity of prizes, or prospects which can be described only by a continuous probability distribution, we cannot obtain an equivalent binary prospect by applying Axiom 3 a finite number of times. The theorem can be proved also in these more general cases, but this requires more sophisticated mathematical tools.

This means that, since the mathematics is beyond doubt, the "validity" of the theorem must depend on the validity of the axioms, i.e., on whether rational decision-makers actually observe these axioms. This question has been discussed at great length in economic literature, so we shall just make a few remarks which may clarify the issue.

Axiom 1 or something equivalent is almost invariably assumed in economic literature. The axiom seems almost trivially true as long as we consider prospects where all outcomes are amounts of dollars. It is, however, less trivial if we consider prospects of non-monetary nature. It is not obvious that there exists a quantity of potatoes which a candidate will accept as adequate compensation for giving up his chances of being elected senator from California.

Axiom 2 is essentially a continuity assumption, of the type commonly made in economic theory. The axiom, as we have formulated it, seems to represent just common sense, and not more than we can assume in a realistic economic theory.

Axiom 3 assumes some sophistication on the part of the decision-maker, or at least that he understands the rudiments of probability theory. The axiom implies that if a decision-maker considers two prospects as equivalent, he will accept to draw lots—at any odds—as to which prospect he will eventually receive.

3.12. The result that a preference ordering over a set of prospects can be represented in the form

$$U\{f\} = \sum u(x)f(x)$$

is often referred to as *The Expected Utility Hypothesis*. In some respects this is a misleading name. Von Neumann and Morgenstern proved the result as a *theorem* derived from some far more basic axioms or "hypotheses." The formula was first proposed by Daniel Bernoulli [1] in 1732 as a hypothesis to explain how rational people would make decisions under uncertainty. The justification Bernoulli gave to support this hypothesis was ingenious, but not quite up to the standard required to convince the economists of this century. Von Neumann and Morgenstern provided an adequate justification for what we in the following shall refer to as the *Bernoulli Principle*, or the Bernoulli decision rule. The theorem was actually proved by Ramsey [10] about 15 years earlier, but no economist seems at that time to have been aware of the significance of the result.

Von Neumann and Morgenstern published their proof as an appendix to the second edition of their book on game theory [9]. This may be the appropriate place for the theorem, but it has had the unfortunate effect of connecting the result with game theory. The Bernoulli Principle is far more fundamental than the other results of game theory, in the sense that it deals with much simpler situations than those game theory is designed to analyze.

Von Neumann and Morgenstern present their proof with an apology that it is "lengthy and may be somewhat tiring for the mathematically untrained reader," and "from the mathematical viewpoint there is the further objection that it cannot be considered deep—the ideas that underlie the proof are essentially very simple" ([9] pp. 617–618).

3.13. The theorem of Von Neumann and Morgenstern implies that in some sense utility becomes "measurable." Many economists have found it difficult to accept this—perhaps not so much because the proof of the theorem was so difficult, but because the preceding generation of economists had waged a fierce and confused battle over the measurability of utility.

Classical economic theory—particularly the Austrian school—considered utility as a measurable property, belonging to any commodity, or to any "market basket" of commodities. The whole theory was built on the concept of *decreasing marginal utility*. In this theory we can find statements which really imply something like

"Three bottles of wine contain only twice as much utility as one bottle."
or
" $2,000 contains only 50% more utility than $1,000."

The outcome of the battle was a general agreement that utility cannot be measured, and that measurement of utility is unnecessary since the whole economic theory can be built on the theory of preference orderings over a set of commodity vectors, or "market baskets," to use the term introduced earlier.

To the victors of this battle, the Von Neumann–Morgenstern revival of measurable utility came as a shock. We get a good idea of the conceptual difficulties which this revival caused for a number of prominent economists from a series of discussion papers [6], [7], [11], and [12] which appeared in *Econometrica* in 1952.

3.14. At the risk of beating a dead dog, we shall briefly indicate how the classical theory can be reconciled with the theory of Von Neumann and Morgenstern. Let us assume that a preference ordering over commodity vectors of the type $\{x_1, x_2, \ldots, x_n\}$ can be represented by a utility function $u(x_1, \ldots, x_n)$.

It was pointed out by Pareto that this preference ordering can also be represented by the utility function $F[u(x_1, \ldots, x_n)]$, where $F(u)$ is an arbitrary increasing function of u.

This means that in analytical work we will have considerable freedom in selecting the utility functions to represent our preference orderings. It is then natural to seek out functions which are easy to manipulate. It is particularly tempting to see if we can find some transformation $F(u)$ of our original utility function u which will make it possible to represent preferences by a function of the simplest possible form, i.e., by a linear function of the type

$$F(u) = a_1x_1 + a_2x_2 + \cdots + a_nx_n.$$

It is easy to see that this class of functions is not rich enough to represent all the preference orderings which we will admit as "rational." Assume, for instance, that a person is indifferent between the two commodity vectors

$$\{x_1, 0, 0, \ldots, 0\} \quad \text{and} \quad \{0, x_2, 0, \ldots, 0\},$$

i.e., between x_1 units of commodity 1 and x_2 units of commodity 2. If his preferences can be represented by a linear function, we must have

$$a_1x_1 = a_2x_2 = \tfrac{1}{2}a_1x_1 + \tfrac{1}{2}a_2x_2.$$

This implies that our person must be indifferent between the three vectors

$$\{x_1, 0, \ldots, 0\}, \{0, x_2, 0, \ldots, 0\}, \text{ and } \{\tfrac{1}{2}x_1, \tfrac{1}{2}x_2, 0, \ldots, 0\}.$$

This is asking too much linearity as a condition for rationality. Few people will question the rationality of a person who states that it is the same to him if he gets two bottles of red wine or two bottles of white wine, but that he would much rather have one bottle of each.

Let us now consider a preference ordering over a set of prospects which can be described by finite discrete probability distributions, i.e., a set of vectors of the type

$$f(x) = \{f(0), f(1), f(2), \ldots, f(M)\}$$

where the sum of the elements is equal to one. It is, of course, possible to represent this preference ordering by a utility function

$$u(f(0), f(1), \ldots, f(M)) = U\{f\}.$$

The preference ordering can also be represented by any function $F(u)$ which increases with u. We have, however, proved in 3.4–3.10 that among the utility functions which can represent a preference ordering over a set of prospects, there will always be one which is a linear expression of a function which we can interpret as the "utility of money."

As an illustration, let us assume that a person is indifferent between the two prospects studied in 3.3.:

$$f_2: \{\tfrac{1}{2}, 0, \tfrac{1}{2}, 0, \ldots\}$$

and

$$f_3: \{\tfrac{1}{4}, \tfrac{3}{4}, 0, 0, \ldots\}$$

It is then natural to assume that he will also be indifferent between all "composite" prospects, or lotteries which will give him f_2 with probability α, and f_3 with probability $1 - \alpha$, i.e., the prospects

$$\alpha f_2 + (1 - \alpha)f_3: \{\tfrac{1}{4}(1 + \alpha), \tfrac{3}{4}(1 - \alpha), \tfrac{1}{2}\alpha, 0, \ldots\}.$$

Hence the linearity which appeared unduly restrictive for commodity vectors seems perfectly natural for probability vectors.

The point we have tried to make here can be expressed as follows: If a person states odd (non-linear) preferences over a set of commodity vectors, we ascribe this to his personal tastes and not to irrationality. If a person states preferences over a set of prospects, he has ample room for personal tastes, but there are certain linearity conditions, inherent in probability theory, which he must observe in order to qualify as rational.

3.15. The Bernoulli Principle is of fundamental importance in economic theory, and a number of authors have improved the proof given by Von Neumann and Morgenstern. One of the first was Marschak [8], who gave an elementary and intuitively very attractive proof. His ideas were followed up by Herstein and Milnor [4], who by sacrificing some of the intuitive appeal were able to give more general, much shorter, and mathematically more elegant proof. These ideas have again been popularized and expanded by Luce and Raiffa ([5] Chap. 2). Debreu [3] has given a very short and elegant proof by using topological methods. The most complete discussion of the theorem has probably been given by Chipman [2] in a paper which makes extensive use of topological methods.

REFERENCES

[1] Bernoulli, D.: "Exposition of a New Theory of the Measurement of Risk," *Econometrica*, 1954, pp. 23–36. Translation of a paper, originally published in Latin in St. Petersburg in 1738.
[2] Chipman, J. S.: "The Foundations of Utility," *Econometrica*, 1960, pp. 193–224.
[3] Debreu, G.: "Cardinal Utility for Even-chance Mixtures of Pairs of Sure Prospects," *Review of Economic Studies*, 1959, pp. 174–177.
[4] Herstein, I. N. and J. Milnor: "An Axiomatic Approach to Measurable Utility," *Econometrica*, 1953, pp. 291–297.
[5] Luce, R. D. and H. Raiffa: *Games and Decisions*, Wiley, 1957.
[6] Malinvaud, E.: "Note on von Neumann-Morgenstern's Strong Independence Axiom," *Econometrica*, 1952, p. 679.
[7] Manne, A. S.: "The Strong Independence Assumption," *Econometrica*, 1952, pp. 665–668.
[8] Marschak, J.: "Rational Behavior, Uncertain Prospects and Measurable Utility," *Econometrica*, 1950, pp. 111–141.
[9] Neumann, J. von and O. Morgenstern: *Theory of Games and Economic Behavior*, 2nd ed. Princeton University Press, 1947.
[10] Ramsey, F. P.: "Truth and Probability," in *The Foundations of Mathematics and Other Logical Essays*, Kegan Paul, 1931.
[11] Samuelson, P. A.: "Probability, Utility and the Independence Axiom," *Econometrica*, 1952, pp. 670–678.
[12] Wold, H.: "Ordinal Preference or Cardinal Utility," *Econometrica*, 1952, pp. 661–663.

Chapter IV

Applications of the Bernoulli Principle
Some Illustrations

4.1. The Bernoulli Principle, which we introduced in Chapter III, gives us a very convenient way of describing economic behavior under uncertainty. In the present chapter we shall illustrate this by discussing some simple applications, and we shall show that the Bernoulli Principle really is the key which opens the door to a general theory for the economics of uncertainty.

The utility function $u(x)$, which represents the preference ordering over a set of prospects, can best be considered as an "operator" which enables the decision-maker—or his computer—to select the best of the available probability distributions. Any decision rule which is consistent in the sense that it satisfies our three axioms can be described in this form. The axioms say nothing about the shape of the function $u(x)$, except that Axiom 2 implies that $u(x)$ must increase with increasing x, i.e., that $u'(x) > 0$.

This means that any particular shape of this function will represent a particular or personal attitude to risk. This naturally leads us to select some simple functions and to study the risk attitudes they represent.

4.2. The simplest possible case is obviously $u(x) = x$. This means that the prospect $f(x)$ is preferred to the prospect $g(x)$ if and only if

$$\sum xf(x) > \sum xg(x).$$

This again means that we consider only the expected value, or the *first moment*, of the probability distributions which describe the available prospect. We prefer the prospect with the greatest expected gain, without paying any attention to the "risk" of deviations from the expected value.

In practice this means that we prefer the prospect of getting $25 if a fair coin falls heads to the prospect of getting $10 with certainty. Some people may actually have preferences like this, but both experience and introspection indicate that there are people who will pick up the $10 bill and walk away rather than engage in a gamble. These people must have an attitude to risk which cannot be represented by a utility function $u(x) = x$.

From our discussion of the St. Petersburg Paradox in 2.9, we can also conclude that no person has preferences which can be represented by a utility function which is linear in the whole interval $(0, \infty)$.

Similar considerations lead us to conclude that $u(x)$ must be *bounded*, an observation first made by Menger [6]. To see this, let us consider an arbitrary small probability $\varepsilon > 0$. If $u(x)$ is unbounded, we can for any x find an $N > x$ so that the inequality

$$u(x) < (1 - \varepsilon)u(0) + \varepsilon u(N)$$

is satisfied. This means that, no matter how long the odds are, we can find a prize which will make the gamble more attractive than any amount of money payable with certainty.

4.3. In order to obtain some results of a more general nature, let us first assume that $u(x)$ is a *concave* function over the interval we consider. This is a natural assumption, since a bounded continuous function is necessarily concave, at least in some intervals.

A function $u(x)$ is concave if the relation

$$u[(1 - p)y + px] \geq (1 - p)u(y) + pu(x)$$

holds for all x and y in the relevant interval, when $0 \leq p \leq 1$. For $y = 0$ the condition reduces to

$$u(px) \geq (1 - p)u(0) + pu(x).$$

The graph of this function will be as indicated by Fig. 3.

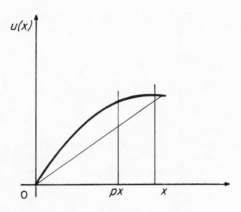

Figure 3

It follows immediately from this condition that if a person whose attitude to risk is represented by $u(x)$ has an amount of money px, he will not stake it in a gamble which will give him

either x with probability p
or 0 with probability $1 - p$.

This is a so-called "fair" gamble, since the stake is equal to the expected gain. It follows in general that if a person's preferences can be represented by a concave utility function—at least in a certain interval—he will not accept a "fair" gamble in this interval.

This person may, however, be interested in insurance, even if this is not quite a fair game. To illustrate this, let us assume that he holds an asset worth x and that there is a probability $1 - p$ that this asset may be lost. This means that he owns a prospect which will give

x with probability p
0 with probability $1 - p$.

If he can insure against this loss, by paying a fair premium $(1 - p)x$, he has the possibility of exchanging his original prospect against a prospect which will give

$$x - (1 - p)x = px \quad \text{with probability 1.}$$

It follows immediately from the definition of concave functions that this degenerate prospect is preferred to the risky one.

4.4. Let us now assume that preferences can be represented by a *convex* utility function, of the shape indicated by Fig. 4, i.e., a function such that

$$u(px) < (1 - p)u(0) + pu(x).$$

In this case our person is willing to gamble, even if the terms are not quite fair.

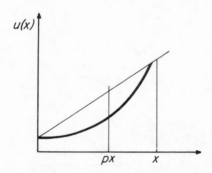

Figure 4

If he holds an amount of cash px, i.e., a prospect with utility $u(px)$, he will be willing to stake his cash on a chance of increasing it to x with probability p, or losing it with probability $1 - p$.

It is usual to say that a person with a concave utility function has a *risk aversion*. He will not by his own choice gamble if odds are fair. He might, however, be induced to gamble if odds are sufficiently favorable to him.

This person is interested in insurance as a means of "reducing the risk" of his original prospect. At the present stage we have to take this term intuitively. We have not defined any measure of risk, and hence it has no meaning to talk about "reducing risk."

Similarly one refers to a person with a convex utility function as a person with *risk preference* or a gambler. Such a person will always accept a bet at fair odds, and he will always be willing to stake his whole fortune on such a bet.

However, it would be too simple if people could be classified either as gamblers or as puritans or, in a different terminology, as "speculators" or "investors." There is some evidence that people have preference orderings which can only be represented by utility functions which are convex in some interval and concave in another.

4.5. This point was first made by Friedman and Savage [5] in a paper which by now is classical. The paper is based on the simple observation that many people who buy insurance to be protected against big losses are at the same time willing to buy lottery tickets. In most cases these people must know that neither of these games is "fair" in the actuarial sense.

If we are going to reconcile these observations with the Bernoulli Principle, we must assume that the utility function has an inflection point near the actual wealth, x_0, of the decision-maker. This means that the graph of the function must be as indicated in Fig. 5. This is in many ways an attractive assumption. It means that an improvement over the status quo has a very high utility, and that a very high loss of utility is assigned to a substantial reduction from the status quo.

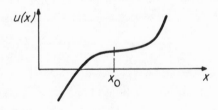

Figure 5

The difficulties occur if we ask what will happen if the person's wealth changes from x_0 to x_1. Will the inflection point remain at x_0 or will it move to x_1? If it moves, will it move instantly, or will the adjustment take some time? To make the problem concrete, we can ask, Will a middle-class person who has fire insurance on his home cancel the insurance if he inherits some hundred thousand dollars?

We shall not discuss this question at present, mainly because it somehow goes outside the legitimate boundaries of our model. The model we have worked with represents a tremendous simplification of the real-life problem. We have, for instance, made no attempt to introduce the *time element*, which obviously must be very important in a realistic model. It seems almost unfair to try to twist our simple model so that it can accommodate the observations of Friedman and Savage. The model was not really designed to satisfy such sophisticated people. To satisfy them, we have to devise a more sophisticated model.

4.6. We have already discussed in 4.2, the simplest case where the utility function was of the form $u(x) = x$. From purely mathematical considerations, a natural next step is to study the function $u(x) = x + ax^2$. This is, however, not a good utility function, since it is not increasing over the whole interval $(-\infty, +\infty)$.

If the prospects we consider can only give finite positive gains, i.e., if we deal only with probability distributions $f(x)$, defined over the interval $(-\infty, M)$, a reasonable preference ordering can be represented by a utility function

$$u(x) = x - ax^2$$

where $0 \le a \le 1/(2M)$. Since $u(x)$ has its maximum for $x = 1/(2a)$, this function will be increasing over the whole interval under consideration.

If we study an insurance company, this may be an adequate working hypothesis. The profits of an insurance company cannot be greater than the amount of premiums received, but there is usually no limit to the losses which the company can suffer. This model has been extensively used by Borch [3].

Let us now assume that preferences can be represented by a utility function $u(x) = x - ax^2$. The utility assigned to a prospect $f(x)$ is then given by

$$U\{f\} = \sum_{-\infty}^{M} (x - ax^2)f(x).$$

We have written M for the upper summation limit as a reminder that the function makes no sense for prospects where infinite gains are possible. We could write ∞ for this limit if we added the condition that $f(x) = 0$ for $x > M$.

From the expression given we find

$$U\{f\} = \sum xf(x) - a \sum x^2 f(x)$$

or

$$U\{f\} = \sum xf(x) - a\left[\sum xf(x)\right]^2$$
$$- a \sum \left[x - \sum xf(x)\right]^2 f(x).$$

We can write this

$$U\{f\} = E - aE^2 - aV$$

where E is the mean and V the variance of $f(x)$.

4.7. The formula in the preceding paragraph has some intuitive appeal. To illustrate this, let us consider a businessman who has heard of neither Bernoulli nor Von Neumann and Morgenstern. Let us assume that this businessman thinks it is a good thing to get a high *expected* profit, but that he thinks some adjustment has to be made for risk. This means that he feels, for instance, that a certain profit of \$1,000 is to be preferred to a 50–50 chance of getting either \$2,000 or nothing.

If this businessman has learned some elementary statistics in his youth, he may think the variance is a good measure of risk. He may then decide to assign the following utility to a prospect $f(x)$:

$$W\{f\} = E - aV$$

and to select the available prospect for which this utility has the highest value. The constant a can be taken as a measure of his "risk aversion."

This is a well-defined decision rule. Our businessman can put the probability distributions which represent the available prospects through a computer and find the prospect which is best according to his attitude to risk.

The decision rule is, however, different from the rule derived from the Bernoulli Principle in 4.6, and this should lead us to suspect trouble. It is easy to see that

$$W\{f\} = E - aV = \sum \left\{ x - a\left(x - \sum xf(x)\right)^2 \right\} f(x).$$

From this it follows that the decision rule of our businessman cannot be represented in the Bernoulli form; i.e., there is no function $u(x)$, independent of $f(x)$, so that

$$W\{f\} = \sum u(x)f(x).$$

This means that the decision rule, in spite of its intuitive appeal, must contain some contradiction in the sense that it violates some of our three axioms.

4.8. To demonstrate that our businessman has been led astray by his commonsense argument, we shall consider a prospect with the outcomes

Gain 1 with probability p

or

Loss x with probability $1 - p$.

For this prospect we find

$$E = p - (1 - p)x = p(1 + x) - x$$
$$V = p + (1 - p)x^2 - [p - (1 - p)x]^2 = p(1 - p)(1 + x)^2.$$

The utility assigned to this prospect will then be

$$W\{f\} = E - aV = p(1 + x) - x - ap(1 - p)(1 + x)^2.$$

Our Axiom 2—and common sense—implies that the utility of the prospect should increase as p increases from 0 to 1. This means that we must have

$$\frac{dW}{dp} = 1 + x - a(1 + x)^2 + 2pa(1 + x)^2 > 0$$

or

$$p > \frac{1}{2} - \frac{1}{2a(1 + x)}.$$

It is obvious that this condition is not satisfied for $p < \frac{1}{2}$ and x sufficiently large.

To drive our point home, let us consider the following two prospects:

Prospect 1:   1 with probability 0.1
                −99 with probability 0.9
Prospect 2:   1 with probability 0.2
                −99 with probability 0.8.

Let us assume that the coefficient of risk aversion $a = 0.1$. It is then easy to verify that our businessman will assign the utility -179 to prospect 1, and hence prefer it to prospect 2, which has a utility of -239. A really unsophisticated observer who has never heard about expectation and variance may well conclude that the basic objective of our businessman is to lose money as fast as he can.

4.9. Conditions of the kind we have considered do not occur if we use decision rules derived from the Bernoulli Principle. If we apply the decision rule of 4.6 to the prospect of the preceding paragraph, we find

$$U\{f\} = E - aE^2 - aV = p\{1 + x + ax^2 - a\} - x - ax^2.$$

This expression will clearly increase with increasing p, provided that a is sufficiently small. We found, however, in 4.6, that this decision rule made sense only if $a \leq 1/(2M)$, where M is the greatest gain obtainable in any of the prospects under consideration. If we apply the rule to prospects which offer gains greater than $1/(2a)$, we will run into contradictions because of the decreasing utility associated with such gains.

To illustrate this point, let us consider a set of binary prospects (p, x), i.e., prospects which will give a gain x with probability p and zero with probability $1 - p$. For such prospects we have

$$E = px$$
$$V = p(1 - p)x^2.$$

The utility of the prospect is, by our rule,

$$U(p, x) = E - aE^2 - aV = px(1 - ax).$$

It is easy to verify that this function will decrease

(i) With increasing p for $x > 1/a$
(ii) With increasing x for $x > 1/(2a)$.

This is, however, against common sense, which clearly requires $U(p, x)$ to be an increasing function of both p and x. An amount $x > 1/a$, payable with certainty, must be better than the certainty of getting nothing, but it gets a lower utility according to our rule.

4.10. A decision rule can obviously be given in many different ways. On the face of it there should be nothing against establishing a rule, i.e., a preference ordering, which depends only on the mean and variance of the prospect. It is then tempting to represent this preference ordering by a utility function of the form $U(E, V)$. This will, however, inevitably lead to contradictions. Only if U is a linear function of

$$E = \sum xf(x)$$

and

$$V + E^2 = \sum x^2 f(x),$$

will a Bernoulli representation be possible. This representation implies, however, that the utility of money is a quadratic function $u(x) = x + ax^2$, which must be decreasing in some interval, and this will, as we have seen, lead to contradictions.

If U is a function of E and V, but not linear in E and $V + E^2$, no Bernoulli representation is possible, and we will run into contradictions of the type discussed in 4.8.

We have discussed this point in some detail since decision rules based on the two first moments have a long tradition in economic literature. There are also good reasons for believing that such rules are used in practice by businessmen who are making decisions under uncertainty. It may, therefore, be useful to stress that such rules can have no general validity. They may work well when applied to a limited class of prospects, but they will inevitably lead to contradictions when applied to *all* possible prospects.

4.11. We shall now consider some additional examples, which will illustrate different aspects of the problem. First we shall consider the rule

Select the prospect with the highest expected gain, provided that the probability of a gain less than A is smaller than a given number α. (A may be negative.) The rule is illustrated by Fig. 6.

In mathematical formulation this means that we select the prospect $f_n(x)$ which gives

$$\max_n \left\{ \int_{-\infty}^{+\infty} xf_n(x) \, dx \right\}$$

subject to

$$\int_{-\infty}^{A} f_n(x) \, dx < \alpha.$$

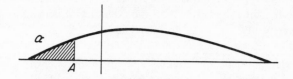

Figure 6

This problem is sometimes referred to as "chance constrained programming."

This rule may sound reasonable, but it implies that a prospect where the area under the probability curve to the left of A is larger than α is always rejected in favor of one where this area is smaller than α, no matter what the shape of the right tail of the curve. For instance the binary prospect

$$A + 1 \text{ cent with probability } 1 + \varepsilon - \alpha$$

or

$$A - 1 \text{ cent with probability } \alpha - \varepsilon$$

will be preferred to the prospect

$$A + \$1,000 \text{ with probability } 1 - \varepsilon - \alpha$$

or

$$A - 1 \text{ cent with probability } \alpha + \varepsilon$$

for any arbitrary small $\varepsilon > 0$.

It is easy to see that there is no function $u(x)$ that will allow this decision rule to be represented in Bernoulli's form. If $\alpha = 0$, such a representation is possible. The condition $\alpha = 0$ means that our decision-maker shies away from any prospect where there is the slightest possibility that the resulting gain will be smaller than A—he considers such prospects as infinitely bad.

This preference ordering can obviously be represented by a utility function

$$u(x) = -\infty \quad \text{for } x < A$$
$$u(x) = x \qquad \text{for } x \geq A.$$

4.12. For a second example we shall go back to the original paper of Bernoulli [2], who starts with the following assumption:

If your wealth increases from x to $x + y$, you will get an increase in utility which is

(i) Proportional to the increase y.

(ii) Inversely proportional to your initial wealth x.

This means that

$$k\frac{y}{x} = u(x + y) - u(x)$$

or

$$\frac{k}{x} = \frac{u(x + y) - u(x)}{y}.$$

Letting $y \to 0$ we obtain

$$u'(x) = k\frac{1}{x}$$

or

$$u(x) = k \log x + c.$$

Bernoulli does not make any assumptions corresponding to our three axioms. His reasoning leads to the conclusion that, if your initial wealth is s, you should prefer the prospect which maximizes the *moral expectation*:

$$\sum f(x) \log (x + s).$$

His arguments are ingenious and seem to take our axioms, or their equivalent, as too obviously true to merit discussion in a learned paper.

Up to the time of Bernoulli it seems to have been considered as self-evident that any rational person would seek to maximize the mathematical expectation when making decisions under uncertainty. This was quite natural, since probability theory originated in the study of gambling situations, where the Law of Large Numbers could be assumed to hold.

Bernoulli's main purpose was to solve the St. Petersburg Paradox. He finds that the certainty equivalent $\bar{x}$ of a St. Petersburg game is determined by the equation

$$\log (s + \bar{x}) = \sum_{n=1}^{\infty} (\tfrac{1}{2})^n \log (s + 2^n)$$

for a person with an initial wealth s.

The assumption of Bernoulli is essentially an axiom which determines the utility function $u(x)$. Taken together with the three axioms in Chapter III, it will give a decision theory without any room for subjective elements. This is proving a little too much, and Bernoulli himself seems to have been aware of this. He is at least willing to accept that a rational person also may compute the moral expectation by the formula

$$\sum (x + s)^{1/2} f(x)$$

4.13. Much of the confusion over utility in economic theory springs from unwarranted extrapolation of the reasoning behind Bernoulli's result. If we take the utility of money as given—to be determined by introspection or from some such psychological principle as that proposed by Bernoulli—it seems very arbitrary to assign the utility

$$u\{f\} = \sum u(x + s)f(x)$$

to the prospect defined by the probability distribution $f(x)$. It is not even obvious that the utility of the prospect can be derived from the utility of money. This has, in fact, been contested by some authors, among whom Allais [1] probably is the most eloquent.

The approach we have taken is different. We have assumed that a complete preference ordering existed over the set of all prospects, and we proved that this preference ordering could under certain conditions be represented by a utility function. An amount of money payable with certainty is a degenerate prospect. This means that a rule which assigns a utility to all prospects must contain a rule for assigning utility to amounts of money. The arguments in classical economic theory over the utility of money are simply bypassed in the approach of Von Neumann and Morgenstern.

4.14. Bernoulli's great contribution was to point out that a rich man and a poor man would not—and should not—decide in the same way when offered an opportunity to choose from a set of prospects. This means that the preference ordering over a set of prospects in general must depend on the initial wealth of the decision-maker.

Economists have often looked for some absolute rule for establishing preference orderings over sets of prospects. This means essentially going back to pre-Bernoullian ideas and seeking a generalization of the classical rule, which took the prospect with the greatest expected value as the best, regardless of initial wealth.

If the desired rule is to satisfy the three rationality axioms of Chapter III, it can be represented by a utility function $u(x)$. If the rule is to be independent of the decision-maker's wealth, the utility functions $u(x)$ and $u(x + s)$ must represent the same preference ordering for all values of s. In 3.10 we proved that this implies that the utility function must satisfy a relation of the form

$$u(x + s) = au(x) + b$$

where a and b are independent of x but may depend on s.

Differentiating this equation with respect to x, we obtain

$$u'(x + s) = au'(x).$$

Differentiation with respect to s gives

$$u'(x + s) = a'u(x) + b'.$$

Combining the two, we obtain the differential equation

$$a'u(x) - au'(x) + b' = 0.$$

If $a' = 0$, the solution of this equation is

$$u(x) = \frac{b'}{a} x + c_1,$$

where c_1 is an arbitrary constant.

If $a' \gtrless 0$, the equation can be written

$$u(x) - \frac{a}{a'} u'(x) + \frac{b'}{a'} = 0,$$

which has the solution

$$u(x) = c_2 e^{-(a'/a)x} - \frac{b'}{a'}$$

with c_2 as an arbitrary constant.

We can express these results as a theorem:

Theorem: If a preference ordering over the set of all prospects (i) *is independent of the actual wealth of the decision-maker, and* (ii) *can be represented in Bernoulli's form, the underlying utility function is either* $u(x) = x$ *or* $u(x) = e^{\alpha x}$.

This theorem is a special case of a more general result due to Pfanzagl [7].

4.15. It is clear that linear transformations of the two utility functions found above also represent preference orderings consistent with our three axioms. For instance, the function

$$u(x) = 1 - ae^{-\alpha x},$$

where $\alpha > 0$, has many attractive properties, and has been used by several writers, among others, Freund [4]. The coefficient α can be taken as a measure of the risk aversion of the decision-maker.

There are, however, no particular reasons why we should ignore Bernoulli's contribution to economic theory and seek a decision rule which is independent of the decision-maker's wealth. Savage has called such decision rules the "rules of the perfect miser," because the decision-maker remains equally greedy, no matter how rich he becomes. If we believe that misers play an important part in economic life, we must, of course, study their decision rules.

It has been suggested that a large corporation may need a decision rule of this kind. In such corporations decision-making must necessarily be delegated to a great number of executives, who cannot always be fully informed about the situation of the corporation. They will then need to have the basic policy of the corporation spelt out in decision rules which are "fool-proof" in the sense that they could not lead an executive astray even if he should not be quite up to date on the actual situation of his corporation. Pfanzagl's theorem implies in fact that only a corporation with a preference ordering which can be represented by one of these two utility functions can really decentralize its organization and delegate the decision-making to representatives who do not consult with each other.

REFERENCES

[1] Allais, M.: "Le comportement de l'homme rationnel devant le risque," *Econometrica*, 1953, pp. 503–546.
[2] Bernoulli, D.: "Exposition of a New Theory of the Measurement of Risk," *Econometrica*, 1954, pp. 23–36.
[3] Borch, K.: "Equilibrium in a Reinsurance Market," *Econometrica*, 1962, pp. 424–444.
[4] Freund, R. J.: "The Introduction of Risk into a Programming Model," *Econometrica*, 1956, pp. 253–263.
[5] Friedman, M. and L. J. Savage: "The Utility Analysis of Choices Involving Risk," *Journal of Political Economy*, 1948, pp. 279–304.
[6] Menger, K.: "Das Unsicherheitsmoment in der Wertlehre," *Zeitschrift für Nationalökonomie*, 1934, pp. 459–485.
[7] Pfanzagl, J.: "A General Theory of Measurement: Application to Utility," *Naval Research Logistics Quarterly*, 1959, pp. 283–294.

Chapter V

Portfolio Selection

5.1. In this chapter we shall study preference orderings over *mixtures* of prospects, or in the terms used in Chapter III, preference orderings over "market baskets" which contain a number of prospects. It is clear that market baskets of this kind are themselves prospects. This means that if we have a preference ordering over the set of *all* prospects, every market basket must have its place in this ordering.

For the sake of simplicity we shall only consider preference orderings which by the Bernoulli Principle can be represented by a polynomial of second degree; i.e., the utility of money is given by a function of the form $u(x) = x + ax^2$. This means that when he is evaluating and ranking prospects, the decision-maker considers only the first two moments (the mean and the variance) of the probability distributions which define the prospects. This simplification can, as we have seen in Chapter IV, lead to absurd results unless we are careful in specifying the set of prospects under consideration.

5.2 Let us now consider two prospects defined by the probability distributions $f(x)$ and $g(y)$. Let the means, or the expected gains, of the two prospects be:

$$E_1 = \sum_x xf(x)$$

$$E_2 = \sum_y yg(y),$$

and let the variances be

$$V_1 = \sum_x (x - E_1)^2 f(x)$$

$$V_2 = \sum_y (y - E_2)^2 g(y).$$

Let us further assume that the stochastic dependence between the two prospects is given by the *joint probability distribution $h(x, y)$*.

By definition we have

$$f(x) = \sum_y h(x, y)$$

$$g(y) = \sum_x h(x, y).$$

As we consider only the first two moments, all relevant properties of the stochastic dependence will be contained in the covariance:

$$C_{12} = \sum_x \sum_y (x - E_1)(y - E_2)h(x, y).$$

5.3. Let us now assume that these two prospects are offered at the same price, which happens to be just equal to the amount of money we have available for investment in an attractive prospect. We can then consult our preference ordering over sets of prospects and buy the prospect which has the higher ranking or, in other words, which has the higher utility.

We can, however, generalize the problem by assuming that we have the opportunity of buying a fraction t of the first prospect and use the remainder of our capital to buy a fraction $1 - t$ of the second prospect. It is easy to see that this transaction will give us a prospect with

Expected return:

$$E = tE_1 + (1 - t)E_2$$

Variance:

$$V = t^2 V_1 + (1 - t)^2 V_2 + 2t(1 - t)C_{12}.$$

If t can take any value between 0 and 1, our problem is to choose the best element from an infinite set of *attainable* prospects. Geometrically this set can be represented by the points on a curve in the EV-plane, as illustrated by Fig. 7.

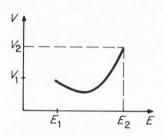

Figure 7

5.4. Our problem is now to select the best of the attainable prospects. This is essentially a problem of determining the *optimal allocation* of our resources. Such problems occur in a great number of economic situations, and the model which we shall develop can be given many different interpretations.

If we interpret the prospects as securities offered in a stock market, the problem will be to determine the optimal *portfolio* which we should buy with our available capital. This interpretation was first suggested by

Markowitz [7] in a delightful little article. Later he expanded the idea into a book [8] which discusses every possible aspect of the problem. This interpretation of the model has become very popular, so we shall refer to our problem as *Portfolio Selection*, the title used by Markowitz for both the article and the book.

5.5. In order to give meaning to the problem, we must introduce a preference ordering over sets of portfolios. We shall do this, and assume that the ordering can be represented by a utility function $U(E, V)$.

This really brings us back to classical economic analysis. The function $U(E, V)$ will determine a family of *indifference* curves in the EV-plane. The solution to our problem is then given by the point where the curve of attainable portfolios touches the indifference curve which represents the highest utility. If, for instance, the utility function is linear, i.e., if

$$U = E - aV,$$

we can write

$$V = \frac{1}{a} E - \frac{1}{a} U.$$

From this we see that the indifference curves are straight, parallel lines and that lines further to the right represent higher utility. It is also worth noting that a steeper slope of the lines corresponds to lower "risk aversion," i.e., to smaller values of the coefficient a.

5.6. Let us now solve the optimizing problem for a linear utility function even if we know that this function implies that we use a decision rule which violates the consistency axioms leading to the Bernoulli Principle. The problem is to determine the value of t which maximizes

$$U = tE_1 - (1 - t)E_2 - a\{t^2 V_1 + (1 - t)^2 V_2 + 2t(1 - t)C_{12}\}.$$

The condition $dU/dt = 0$ gives

$$t = \frac{E_1 - E_2 + 2aV_2 - 2aC}{2a(V_1 + V_2 - 2C)}.$$

With our formulation of the problem, this solution has a meaning only if $0 \leq t \leq 1$. However, this restriction is not really necessary. If our formula gives a solution $t < 0$ or $t > 1$, we can take this to mean that the optimal action is to go short in one security and take a corresponding long position in the other.

5.7. To generalize this simple result, let us consider a stock market with n different securities. Let

E_i = expected return on security i;

V_i = the variance of the return on security i;

C_{ij} = the covariance of the returns on securities i and j;

t_i = the fraction of our capital which we invest in security i.

It is easy to see that this investment decision will give us a portfolio with
Expected return:

$$E = \sum_{i=1}^{n} t_i E_i$$

Variance:

$$V = \sum_{i=1}^{n} t_i^2 V_i + 2 \sum_{i \neq j}^{n} C_{ij} t_i t_j.$$

The latter expression can be written in the more symmetric form

$$V = \sum_{i=1}^{n} \sum_{j=1}^{n} C_{ij} t_i t_j$$

if we write $V_i = C_{ii}$.

Our problem is then to determine an *allocation* of our capital, described
by a vector $(t_1, \ldots, t_n)$, which maximizes $U(E, V)$. Usually there will be
some restraints on this problem.

If no security can be "sold short," we must have

$$t_i \geq 0 \quad (i = 1, \ldots, n).$$

If we cannot "buy on margin," i.e., if there are no credit facilities, we have

$$\sum_{i=1}^{n} t_i \leq 1.$$

If our whole capital must be invested in these n securities, the equality sign
will apply in the latter condition.

If our decision rule satisfies the consistency conditions which lead to the
Bernoulli Principle, $U(E, V)$ will be a quadratic function in $t_1, \ldots, t_n$.
This means that we can solve our problem by the familiar techniques of
quadratic programming.

5.8. Markowitz takes a different approach, apparently to avoid the
Bernoulli Principle, which he considers controversial. His only assumption
is that there exists a utility function $U(E, V)$ which satisfies the two
conditions:

$$\frac{\partial U}{\partial E} > 0 \quad \text{and} \quad \frac{\partial U}{\partial V} < 0.$$

This means that, *other things being equal*:

(i) A greater expected return is preferred to a smaller one.
(ii) A smaller risk (measured by the variance) is preferred to a larger one.

He then picks some value $\bar{E}$ = *required return* and determines the portfolio which has the smallest variance among those which give an expected
return equal to $\bar{E}$.

Mathematically this leads to the problem of minimizing

$$\sum_{j=1}^{n} \sum_{i=1}^{n} t_i t_j C_{ij}$$

subject to the constraints

$$\sum_{i=1}^{n} t_i E_i = \bar{E}$$

$$\sum_{i=1}^{n} t_i \leq 1 \qquad t_i \geq 0.$$

The last two constraints must of course be modified if short selling and margin buying are permitted.

This is again a problem in quadratic programming. The solution to the problem will be a vector $\{t_1, \ldots, t_n\}$ which will determine the best or the *efficient portfolio* with expected return equal to $\bar{E}$.

5.9. If this problem is solved for a number of different values of $\bar{E}$, we will obtain a set of efficient portfolios. Each of these portfolios can be represented by a point in the EV-plane, and these points will lie on a curve as indicated by Fig. 8. Points above this curve may be attainable,

Figure 8

but they will not be efficient. Points below the curve will represent portfolios which are not attainable. The curve is drawn through the origin. This implies that there exists one riskless security—for instance cash—which earns no return. In some cases it may be natural to assume that there exists a riskless investment which earns a positive return E_0—for instance government bonds or saving bank deposits. The curve representing the efficient portfolios will then intersect the abscissa axis in the point $(E_0, 0)$.

The basic idea behind the Markowitz model is that the investor should gaze at this curve for a sufficiently long time and then decide on a point which, "more than any other, satisfies his needs and preferences with respect to risk and return" ([8] p. 23).

5.10. The real justification of the Markowitz approach must be that people find it easier to compute than to state a complete and consistent set of preferences. This is not an unreasonable assumption, and it may make good sense to split up the problem and reshuffle the different elements. We can then state the obvious and uncontroversial part of the preference ordering, and as a next step compute all efficient portfolios, even if only one of them will be chosen. This means that we reduce the decision problem to its simplest possible form, in the way discussed in 2.12, and then make the final choice by applying a criterion which clearly must be "subjective."

The alternative, the more orthodox approach of 5.7, will oblige the investor to state preferences over all points in the EV-plane, even if some of them do not represent attainable portfolios. This can be done by specifying one single number—the "risk aversion" a—if the investor accepts the consistency conditions of the Bernoulli Principle and decides to consider only the mean and the variance of the prospects. However, many decision-makers seem to consider this a formidable problem.

The division of labor implied by the Markowitz model may correspond fairly well to actual business practice. The usual procedure seems to be that the technicians sort out the problem, if necessary with the help of a computer, and present their findings to top management. These findings will then be considered behind the closed doors of the Board Room, and a decision will be made.

The alternative procedure would be for top management to spell out their preferences—or the objectives of the corporation—and leave the actual decision to the technicians—or to the computer.

It is not difficult to think of good reasons for preferring the first procedure to the second. The company may, for instance, gain some competitive advantages by keeping its objectives secret. Secrecy may also save top management from the embarrassment of having technicians pointing out that the stated objectives are inconsistent.

5.11. In general it is not clear whether it is best to compute first and think afterward, or vice versa. It is, however, not surprising that IBM has shown a considerable interest in the Markowitz model and has tried to adapt it for practical use. To illustrate how this is supposed to be done, it may be useful to quote a few passages from a recent IBM publication [5]:

"Program input consists of estimates in the form of probability beliefs on the expected return and possible risk associated with individual securities, along with data on the correlation of their price movements in the market, as they are compiled by professional security analysts."

"Within the computer, data is subjected to a rigorous optimization by a mathematical model."

"The portfolio manager then applies his experience and judgement to this quantitative information, and establishes an investment strategy for the fund or client."

This may sound quite well, except that the passages we have quoted tell us neither how we should obtain the input nor how we should make our decision on the basis of the output.

From our discussion in the preceding chapters, it is clear that the rule which leads to the final decision must contain some subjective elements, which in a sense are beyond the scope of normative analysis. We shall, however, leave this question for the time being and study the input which we will need to make sensible use of the model.

5.12. To attack the problem we must first specify what we mean by the "return" on a security. For our purpose, return is a stochastic variable, and it must take account of both dividend payments and capital appreciation. To illustrate the point, Markowitz begins his book by discussing a simple example. He selects nine securities quoted on the New York Stock Exchange, and for each of the 18 years in the period 1937–1954 he calculates the return according to the formula

Return = {Price at the end of the year
— Price at the beginning of the year
+ Dividends paid during the year}
Divided by: {Price at the beginning of the year}

TABLE 2

Year	Return	
	American Tobacco	American Tel. & Tel.
1937	−0.305	−0.173
1938	0.513	0.098
1939	0.055	0.200
1940	−0.126	0.030
1941	−0.280	−0.183
1942	−0.003	0.067
1943	0.428	0.300
1944	0.192	0.103
1945	0.446	0.216
1946	−0.088	−0.046
1947	−0.127	−0.071
1948	−0.015	0.056
1949	0.305	0.038
1950	−0.096	0.089
1951	0.016	0.090
1952	0.128	0.083
1953	−0.010	0.035
1954	0.154	0.176

His findings for two of the securities are reproduced in Table 2. From these data he finds

For American Tobacco

Average return:	0.066
Variance:	0.0533
Standard deviation:	0.231

For American Tel. & Tel.

Average return:	0.062
Variance:	0.0146
Standard deviation:	0.121

The correlation coefficient between the returns on these two securities is

$$r_{12} = 0.77.$$

This is the highest correlation coefficient found for any pair of the nine securities considered. The lowest correlation coefficient is 0.18, found for *Coca Cola* and *Atchison, Topeka & Santa Fe.*

The fact that all correlation coefficients turn out to be positive may have some significance. This seems to indicate that the returns on all securities move together—up or down—with the market as a whole. If this is true, it means that it is impossible, even for the most careful investor, to protect himself against the adverse effects of general fluctuation on the stock market. This conclusion seems very plausible for intuitive reasons, and it has been confirmed by a number of statistical studies, i.e., by a paper by Borch [1].

5.13. From his analysis of the performance of nine securities, Markowitz obtains

9 expected (or average) returns
9 variances
36 covariances

The 54 numbers express in a condensed manner the essential information about how the nine securities behaved during the period 1937–1954, and that is all. An investor who intends to use the Markowitz model will need the 54 corresponding numbers representing the *future* performance of the securities. For this purpose, observations of past performance are probably relevant, but it is by no means clear how this information should be used to form estimates or guesses about the future.

The simplest approach is obviously to assume that securities will behave in the future as they have behaved in the more recent past. It is, however, possible to find good arguments for the opposite assumption. Stock market experts will frequently make statements to the effect that a security will enter a period of rest or consolidation after a period of rapid growth or violent fluctuation.

5.14. The Markowitz model is a procedure for making systematic use of our probabilistic beliefs about the future performance of securities. It is, however, not clear how these beliefs are to be obtained—for instance by a systematic analysis of past performance. It may, therefore, be useful to make a digression and refer to the statistical evidence that prices in a stock market behave in an almost completely random fashion. This should indicate that it is impossible to predict future price movements by merely studying statistics of past price behavior, and hence that "chartists" and "trend watchers" are wasting their time—and possibly their own or other people's money.

The classical paper in this field was published by Cowles [3] in 1933. In this paper he showed that by buying and selling securities at random one would on the average do just as well as one would by following the advice of professional—and expensive—investment consultants.

Cowles himself and other authors have followed up these results in further studies, based on more extensive statistical data. A number of these studies have been collected and republished in a book edited by Cootner [2]. Cowles' findings have quite naturally been ignored by security analysts. This profession seems to be doing very well—as astrologers once did—apparently because the public firmly believes that there must be some law behind price movements in the market. The person who knows or can guess this law should be able to make money by playing the market.

Kendall [6], whose findings confirmed those of Cowles, comments as follows: "Investors can, perhaps, make money on the Stock Exchange, but not, apparently, by watching price-movements and coming in on what looks like a good thing." However, he adds, a little sadly, "But it is unlikely that anything I say or demonstrate will destroy the illusion that the outside investor can make money by playing the markets, so let us leave him to his own devices."

The findings of these authors seem to imply that the behavior of security prices in the past does not contain any information which can help us to predict future price movements. This should not surprise an economist. If it were possible by statistical analysis to pick out one security as the obviously best buy in the market, we would expect the price of this security to rise, up to the point where the security in question was no longer the obvious buy.

5.15. To elaborate the last point, let us assume that we try to apply the Markowitz model. We may then well find that some securities do not enter into any efficient portfolio. This obviously means that, according to the personal beliefs which we fed into the computer, these securities are in some sense "over-priced." If they were offered at a lower price, expected return would presumably be higher, and these neglected securities would find their place in some efficient portfolio.

It is, however, evident that some people are willing to hold and buy these securities which, according to our beliefs, no rational person would touch regardless of what risks he is willing to take. We can, of course, assume that these people are irrational and try to find a way to profit by their stupidity. It may, however, be safer to take a more tolerant attitude and assume that these people act rationally, but on beliefs which are different from our own, though just as sound.

In this situation we can stick to our own beliefs and act on them alone. It is, however, not certain that we will have the "courage of our convictions" if we learn that insurance companies and mutual funds are buying the securities which we think should be excluded from any efficient portfolio. It may be very tempting to adjust our beliefs to those of other "men of experience and sound judgement." If we do this, we will admit more securities into the "efficient" portfolio. If we carry this adjustment process to the extreme, we may arrive at the conclusion that all securities are equally good at the current market prices; i.e., we can save the computing and buy at random. This gives us a theoretical explanation of the observations made by Cowles and his followers.

5.16. It is obvious that the Markowitz model represents a tremendous simplification of the real-life investment problem. However, in all its attractive simplicity the model seems to capture some of the essential elements of the real problem, so it appears very desirable to test the model against observations of economic behavior.

This question has been taken up by several authors. We shall, however, only discuss the work of Farrar [4], who investigated some *Mutual Funds* to find out if their investment behavior could be explained by the Markowitz-model. Farrar studied the two questions

 (i) Are the portfolios actually held by mutual funds efficient in the sense of the Markowitz model?
 (ii) If a portfolio is approximately "efficient," does it represent an attitude to risk which is consistent with the stated investment objectives of the fund?

A portfolio can be described as efficient only with reference to a particular set of beliefs about future return, expressed in terms of expectations and covariances. The actual beliefs of the mutual funds are neither stated nor known, so it is impossible to verify whether the portfolios held are efficient with respect to these unknown beliefs. Hence a test of the Markowitz model must imply a test of some hypothesis as to how beliefs about future returns are formed.

In this context it may be worth noting that it is impossible to test whether a person will do better than average if he uses the Markowitz–IBM approach to his investment problem. If the beliefs of this person are correct, he can be expected to do best if he tries to maximize expected return. This is nothing but a tautology. If this person diversifies his investments, his expected return will be smaller, but he will run less risk. This may give him some peace of mind, which he may consider an adequate compensation for the reduction in expected return. It is obviously impossible to test statistically whether our person has sacrificed just the right amount of expected return.

From statistical studies we may, however, learn something about the behavior of certain groups of investors. This knowledge will have a "practical" value if we can use it to make profits for ourselves.

5.17. Farrar starts out by computing the efficient portfolios available in the market under the assumption that the behavior pattern of returns observed in the past will repeat itself in the future. It is obviously impossible to carry out these computations for all securities in the market. In order to simplify the problem, Farrar takes average return for 58 different classes of securities, listed in Standard and Poor's *Trade and Security Statistics*. He does this for each month of the period January 1946–September 1956 and computes the expectations and the covariance matrix as we did in 5.12.

It would be possible, but not very convenient, to work with this 58×58 matrix and determine the efficient portfolios. These portfolios would represent efficient allocation of capital, not among individual securities, but among classes of securities, i.e., different industries, preferred or common stock, bonds, etc. However, Farrar manages to reduce the problem to an 11×11 matrix by a clever application of *factor analysis*. The idea behind this technique is the following:

Assume that we have n strongly interdependent stochastic variables z_1, $z_2, \ldots, z_n$, which may represent return in n industries. It may then be possible to find a few, say three, stochastically independent variables x_1, x_2, and x_3, so that equations of the form

$$z_1 = a_1x_1 + b_1x_2 + c_1x_3$$

$$z_2 = a_2x_1 + b_2x_2 + c_2x_3$$

$$\cdot \quad \cdot \quad \cdot \quad \cdot \quad \cdot \quad \cdot \quad \cdot \quad \cdot$$

$$z_n = a_nx_1 + b_nx_2 + c_nx_3$$

hold with sufficiently good approximation. Here a_i, b_i, and c_i are constants to be determined so that we in some sense obtain the best possible fit.

To represent a number of dependent variables by a smaller number of independent variables can be taken as a purely statistical problem. In many cases it may, however, be possible to give a concrete interpretation to the components or the "factors" x_1, x_2, and x_3. If, for instance, these stand for return on investment in the three industries

1. Automobile production
2. Shipbuilding
3. Aluminum production

it is quite reasonable to assume that return on investment in the steel industry may be determined approximately by an expression such as

$$z = 0.8x_1 + 0.3x_2 - 0.1x_3.$$

5.18. From his 11×11 covariance matrix, Farrar computes the efficient portfolio under each of the following three assumptions about future beliefs:

(i) Dividends and prices will vary as in the past.
(ii) Dividends will vary as in the past, and prices will follow the linear trend fitted to the changes over the last 12 months.
(iii) Dividends will vary as in the past, and prices will follow the exponential trend of the last 12 months (i.e., linear extrapolation of the current growth rates).

He finds that these different assumptions do not lead to significant differences in the composition of the efficient portfolios.

As a next step Farrar computes the expected return and the variance for the portfolios actually held by 23 mutual funds. He then finds that all these portfolios correspond to points in the EV-plane which lie fairly close to the curve which represents the efficient portfolio.

This is a very pretty result, and it obviously has some significance. It is, however, impossible to evaluate the result properly without engaging in a statistical analysis far more extensive than Farrar's study.

Farrar has not, and does not claim to have, proved that managers of mutual funds are clever. He may, however, have proved that these managers act on the belief that the future performance of a security, or a class of securities, is to a very large extent determined by behavior in the past. This may be taken to mean that security analysts are more statisticians than economists, in the sense that they predict by extrapolation and charts rather than by economic theory.

5.19. In 5.5 we found that if the preferences of an investor could be represented by a utility function of the form

$$U(E, V) = E - aV$$

his indifference curves in the EV-plane would be the parallel lines

$$V = \frac{1}{a} E - \frac{1}{a} U.$$

This means that if the investor selects a point on the curve representing the efficient portfolios, the tangent to the curve at this point will have the slope $1/a$ (see Fig. 8). The number a can, as we have seen, be interpreted as a measure of the "risk aversion" of the investor.

An investor, for instance, with a high risk aversion will select a point far to the left on the curve, i.e., a point where the tangent is almost horizontal. By observing the point actually selected, we can then determine the risk aversion which would lead the fund to this selection.

In its prospectus and advertisements a mutual fund will usually make some statements about its attitude toward risk. It is then possible to check whether these statements are consistent with the portfolio actually held by the fund.

5.20. In order to study this question, Farrar divides his 23 mutual funds into the following three classes, according to their stated objectives:

1. *Balanced funds*, i.e., funds which aim at a stable return. These funds state more or less explicitly that they intend to sacrifice something in order to avoid fluctuations in the return.
2. *Stock funds*, i.e., funds which are prepared to accept the fluctuations normally associated with investment in good common stock.
3. *Growth stock funds*, i.e., funds which openly state that they are "speculative," or that they will take considerable risks in order to make important long-term gains.

The coefficient of risk aversion found for the 23 funds, is given in Table 3.

5.21. We can draw a number of interesting conclusions or conjectures from Table 3. It is remarkable that all the ten "balanced funds" have portfolios with almost identical *EV*-properties. The exception is the National Balanced Series, which appears far more conservative than any of the other funds.

In the class of "stock funds" there is a wider spread, as we would expect. It appears, however, that Axe-Houghton manages its "stock fund" more conservatively than we would expect from the stated objectives of the company.

The six "growth stock funds" also seem to hold portfolios with very similar *EV*-properties.

It is particularly interesting to look at the investment companies which manage several funds, designed to meet the needs of different classes of investors. It appears that the National Series really offers an investor different alternatives. On the other hand, companies such as Axe-Houghton, Keystone, and Massachusetts Investor offer funds with different objectives, but they hold portfolios with substantially the same *EV*-properties.

TABLE 3

Fund	Coefficient of risk aversion
Balanced funds	
National Balanced Series	48
Axe-Houghton A	38
Affiliated Fund	34
Commonwealth Investment Co.	34
Eaton & Howard Balance Fund	34
Investor's Mutual	34
Knickerbocker (Bal.) Fund	34
Axe-Houghton B	33
Diversified Investment Fund	33
Institutional Foundation Fund	32
Stock funds	
Axe-Houghton Stock Fund	31
Eaton & Howard Stock Fund	22
Investor's Stock Fund	22
Institutional Income Fund	12
National Stock Series	11
Mass. Investor's Trust	10
Keystone S-1 (Stock)	6
Growth stock funds	
Mass. Investor's Growth Fund	10
Institutional Growth Fund	9
Knickerbocker Capital Venture	8
Diversified Growth Stock	7
Keystone S-3 (Speculative)	7
National Growth Stock Series	6

5.22. It is obvious that we cannot attach much significance to these conclusions. It seems, however, that the Markowitz model captures some of the essential elements of the real-life investment problem as seen by professional investors. This is a little surprising when we consider the basic simplicity of the model, and it appears even more surprising when we recall the heroic simplifications which Farrar has to make in order to carry through his statistical analysis.

The Markowitz model is "timeless" in the sense that it cannot distinguish between short-term and long-term investment objectives, and most professional investors will insist that this distinction is very important in practice.

Further, the model is built on the *EV* rule, which in itself contains a contradiction, as we demonstrated in 4.8. We can avoid this contradiction by restricting the class of prospects under consideration. Even then the rule does not appear very realistic by introspection. The rule ignores all kinds of "skewness" in returns, and this implies, for example, that we should be indifferent between the following four prospects:

(i) A 50–50 chance of getting either 0 or 2.

(ii) A prospect which will give a return of
either -0.997 with probability 0.998
or 999 with probability 0.002.

(iii) A prospect in which the probability of a return x is given by

$$f(x) = e^{-1} \frac{1}{x!}, \ (x = 0, 1, 2, \ldots)$$

(iv) A prospect in which the probability density of a return x is given by

$$f(x) = \frac{1}{(2\pi)^{\frac{1}{2}}} e^{-\frac{1}{2}(x-1)^2}$$

Some persons may actually consider these four prospects as equivalent, since they all have mean and variance equal to 1. However, if we want to save the Markowitz model, it may be more realistic to assume that the stock market does not offer prospects which are so dramatically different.

REFERENCES

[1] Borch, K.: "Price Movements in the Stock Market," *Skandinavisk Aktuarietidskrift*, 1964, pp. 41–50.
[2] Cootner, P. H. (ed.): *The Random Character of Stock Market Prices*, The MIT Press, 1964.
[3] Cowles, A.: "Can Stock Market Forecasters Forecast?" *Econometrica*, 1933, pp. 309–324.
[4] Farrar, D. E.: *The Investment Decision under Uncertainty*, Prentice-Hall, 1962.
[5] IBM: "Portfolio Selection: A New Mathematical Approach to Investment Planning," *IBM General Information Manual*, E–20-8107, 1961.
[6] Kendall, M. G.: "The Analysis of Economic Time Series," *Journal of the Royal Statistical Society, Series A*, 1953, pp. 11–25.
[7] Markowitz, H. M.: "Portfolio Selection," *The Journal of Finance*, 1952, pp. 77–91.
[8] Markowitz, H. M.: *Portfolio Selection—Efficient Diversification of Investments*, Wiley, 1959.

Chapter VI

The Bernoulli Principle—
Observations and Experiments

6.1. In Chapter III we showed how the Bernoulli Principle could be derived as a theorem from three simple axioms. These axioms appear very plausible. It seems almost self-evident that a rational person who seeks a formal rule for economic decisions under uncertainty cannot possibly accept a rule which violates any of the three axioms.

In Chapters IV and V we tried to demonstrate how extremely useful the Bernoulli Principle is. We showed that a number of important economic problems could be formulated and analyzed with great ease by the application of this principle. We were, however, able to say little about the shape of the utility function which represents the decision rule.

This naturally leads us to seek the answers to the following two questions:

(i) Do the axioms hold in practice, i.e., are they observed by important groups of people who make decisions under uncertainty?
(ii) If the axioms hold, what is the shape of the utility function representing the preference ordering of a typical decision-maker in the different situations which we want to study?

Since introspection cannot carry us much further, it is natural to resort to observations of economic behavior or, if possible and necessary, to controlled experiments, i.e., to take an empirical approach.

6.2. To illustrate the first point, we shall discuss an example due to Allais [2]. He considers the following two situations:

Situation 1: We have to choose between the prospects A and B.

A will give a gain of \$1 million with certainty.
B will give
    Either \$5 million with probability 0.10
    or \$1 million with probability 0.89
    or nothing with probability 0.01.

Most people seem to prefer prospect A, i.e., prefer to take the million rather than the risk of getting nothing.

If this decision is based on a preference ordering which satisfies our three axioms, there must exist a function $u(x)$ such that

$$u(1) > 0.1\, u(5) + 0.89\, u(1) + 0.01\, u(0).$$

Situation 2: We have to choose between the prospects C and D.

C will give
   Either \$1 million with probability 0.11
   or nothing with probability 0.89.
D will give
   Either \$5 million with probability 0.1
   or nothing with probability 0.9.

In this situation most people seem to prefer D.
   If this decision is based on the same preference ordering, we must have

$$0.1\ u(5) + 0.9\ u(0) > 0.11\ u(1) + 0.89\ u(0).$$

If we add the two inequalities, we obtain

$$0.1\ u(5) + u(1) + 0.9\ u(0) > 0.1\ u(5) + u(1) + 0.9\ u(0).$$

This is obviously a contradiction, since we have assumed strict preferences and strict inequalities.
   If in situation 2 we had chosen C, a similar argument will give

$$0.11\ u(1) > 0.1\ u(5) + 0.01\ u(0).$$

It is easy to see that we can find a function $u(x)$ which satisfies this condition, i.e., which can be interpreted as the utility function representing the preference ordering of the decision-maker.

6.3. The example of Allais illustrates two points:

(i) If we observe that a person has chosen A in situation 1, we can *predict* that he will choose C in situation 2, assuming that he makes his decisions in a rational manner.

We can test our prediction by making further observations or by arranging an experiment which places our person in situation 2.

(ii) If a person in situation 1 has chosen A, he must choose C in situation 2 if he wants to be consistent, i.e., the first choice *commits* him.

In 1952 Allais prepared a number of examples of this kind, and asked several prominent economists how they would choose in such situations. The questionnaire circulated by Allais has been published [1], but not the systematic analysis of the replies. It is, however, generally known that leading economists made choices which implied an inconsistent preference ordering.
   One of them was Savage, who admits that he was trapped by the example we have quoted, and that he chose A and D. He adds, however, that when the contradiction was pointed out to him, he reconsidered the problem and reversed his choice in situation 2 from D to C. Savage states that when he

did this, he felt that he *corrected an error* ([10] p. 103). This seems to mean that Savage feels a certain obligation to make his decisions in a consistent manner—a feeling which may not be shared by other people.

Savage justifies his "second thoughts" by the following argument: The decision problem can be realized by a lottery with 100 numbered tickets, with prizes as indicated by Table 4. If one of the numbers 12 through 100

TABLE 4
Prizes in million dollars

		Number drawn		
		1	2–11	12–100
Situation 1	A	1	1	1
	B	0	5	1
Situation 2	C	1	1	0
	D	0	5	0

is drawn, it does not matter in either situation which prospect is chosen. If one of the numbers 1 through 11 is drawn, the two situations are identical. The real problem is then to decide whether one prefers an outright gift of $1 million, or a 10-to-1 chance of winning $5 million. This means that the logical choices are either A and C, or B and D.

6.4. Some people may not feel Savage's compulsion to be consistent. If a person insists that he prefers A in situation 1 and D in situation 2, there is little we can do. Preferences are necessarily subjective, and there is no point in arguing. We can note that if our person has some general rule for making decisions which leads to these choices in the two situations, this rule cannot satisfy our three axioms, i.e., there are good reasons for calling the rule inconsistent. We may, however, go one step further and let him choose either $A(\alpha)$ or $B(\alpha)$ in the following situation:

Situation α
$A(\alpha)$ will give
   Either $1 million with probability $1 - \alpha$
   or nothing with probability α.
$B(\alpha)$ will give
   Either $5 million with probability 0.10
   or $1 million with probability $0.89 - \alpha$
   or nothing with probability $0.01 + \alpha$.

Our person has already stated that he prefers

$$A(\alpha) \text{ to } B(\alpha) \quad \text{for } \alpha = 0$$

and

$$B(\alpha) \text{ to } A(\alpha) \quad \text{for } \alpha = 0.89.$$

We can then ask him to specify the value of α for which he will reverse his preferences. It is then likely that he will either admit his inconsistency or insist that there is something magic about the number 0, i.e., that a zero-probability is radically different from any positive probability, no matter how small. This clearly means that he rejects the continuity assumptions inherent in our axioms.

If the axioms are satisfied, so that a utility function exists, the statement "$A(\alpha)$ preferred to $B(\alpha)$" implies

$$(1 - \alpha)u(1) + \alpha u(0) > 0.10\, u(5) + (0.89 - \alpha)u(1) + (0.01 + \alpha)u(0)$$

or

$$0.11\, u(1) > 0.01\, u(5) + 0.01\, u(0).$$

This relation is independent of α, so that preferences cannot be reversed when α varies.

Samuelson, when discussing these axioms, recalls a story about an old farmer who considered everyone in the world crazy except himself and his wife. This farmer used to add "sometimes I am not quite certain about her." Samuelson is tempted to assume that only he himself and Professor Savage are so rational that they will always observe the axioms, but he adds "sometimes I am not quite certain about myself" ([9] p. 678).

6.5. So far we have taken it for almost self-evident that in situation 1 "ordinary" rational people would prefer A to B. However, is this—on second thought—really so obvious? After all what would an ordinary person do with a million dollars? He would not, and probably could not, just spend it. It is likely that he would invest most of the money, and that means that he really would be exchanging the prospect A for a prospect more like B.

If, for instance, the person who selected A would spend $10,000 and invest the rest in growth stock, he would in reality be changing A for a prospect $\bar{B}$ of the type:

$3 million with probability 0.10
$1 million with probability 0.89
$10,000 with probability 0.01.

The point we want to make is that many people may say they prefer A to B, and defend their choice. Having done this, they may well make decisions which imply that they prefer $\bar{B}$ to A.

It is doubtful if examples of this kind can contribute much to our knowledge about economic behavior under uncertainty. Most people are not used to tossing coins or throwing dice for millions of dollars, and one should probably not attach very much significance to their statements as to how they would make decisions in such situations. One should at least admit that rational people may well make "mistakes" when they state how they would decide in situations which they have never had to consider seriously.

These considerations may lead us to study the decisions people make in real life, when they stand to gain or lose, if not millions, at least thousands of dollars. In such situations we can assume that the people take the problem seriously, and presumably also that they try to behave in a consistent manner. We shall see later in this chapter that such studies form an important source of information about economic behavior under uncertainty. They are, however, of minor value in the present context, since a decision-maker in real life usually will have only an incomplete knowledge of the probabilities involved.

6.6. It is not easy to find a decision problem which satisfies the two following conditions:

(i) The decision-maker takes the problem seriously, because he gets a real payoff.
(ii) The relevant probabilities are known, both to the decision-maker and the observer.

When we cannot find such problems, it is natural that we should try to construct them, and this leads us to arrange *controlled experiments* to gain information about the decision rules of the people we are interested in.

Before we discuss experimental work, it is necessary to say a few words about *subjective probabilities*, a concept which we shall study in more detail in Chapter XIV.

Let us assume that we ask a person to choose between the prospects A and B:

A will give a gain x_1 if the event E_1 occurs.
B will give a gain x_2 if the event E_2 occurs.

Let us further assume that the preferences of our person can be represented by a utility function $u(x)$ of unknown form, and let

$$\text{Prob}\,\{E_1\} = p_1 \quad \text{and} \quad \text{Prob}\,\{E_2\} = p_2.$$

If for some reason our person prefers A to B, the Bernoulli Principle implies that

$$p_1 u(x_1) > p_2 u(x_2),$$

where for the sake of simplicity we have assumed $u(0) = 0$.

If we consider the probabilities as known, this choice gives us some information about the utility which represents the underlying preference ordering. If, on the other hand, we take the utility function $u(x)$ as known, the observed choice gives us some information about the *subjective probabilities assigned to the events E_1 and E_2*.

If, for instance, $x_1 = x_2$, we conclude from

$$A \text{ preferred to } B$$

that

$$\text{Prob } \{E_1\} > \text{Prob } \{E_2\}$$

at least in the mind of the person who made the choice.

Now let

$E_1 = $ A fair coin will fall heads.

$E_2 = $ Our university team will win its next football game.

$x_1 = \$10.$

$x_2 = \$100.$

Let us further assume that the person we study knows something about probability, and that he really believes that

$$\text{Prob } \{E_1\} = \tfrac{1}{2}.$$

If he prefers A to B, we have

$$\tfrac{1}{2} u(10) > \text{Prob } \{E_2\} u(100)$$

or

$$\text{Prob } \{E_2\} < \frac{u(10)}{2u(100)}.$$

From this we can obviously conclude that our person believes that

$$\text{Prob } \{E_2\} < \tfrac{1}{2}$$

i.e., that the odds are that our team will lose.

We can further conclude

(i) Either he thinks there is a very low probability that our team shall win

(ii) or his utility function is very "flat."

In general, we can explain observed choices either by assuming that the utility function has a particular shape, or by making assumptions about how our person forms subjective probabilities or "expectations" on the basis of the information available to him. The Bernoulli Principle makes it possible to separate the two elements in the decision problem.

6.7. The first attempt to measure utility by controlled experiments was made in 1950 by Mosteller and Nogee [8], who studied a group of Harvard undergraduates and some members of the Massachusetts National Guard. Mosteller and Nogee found that their subjects were not perfectly consistent in their choices, but that, in spite of this, the theory based on the Bernoulli

Principle had a considerable predictive power. They also found that the utility function of the Harvard students differed significantly from utility functions which represented the preferences of the guardsmen.

We shall not discuss this experiment in further detail. Instead we shall give a brief account of another experiment, conducted by Davidson, Suppes, and Siegel [5]. This may give some useful indications about the problems we encounter in "experimental economics"—problems which may be unfamiliar to most economists, since economics is not usually considered an experimental science.

6.8. The experiment of Davidson, Suppes, and Siegel consisted essentially in asking the subjects to bet either "heads" or "tails" in situations like the following:

If you bet *heads*, you will
(i) gain 5 cents if right
(ii) lose 5 cents if wrong.

If you bet *tails*, you will
(i) gain 6 cents if right
(ii) lose 5 cents if wrong.

The situation can be represented by the following "payoff matrix":

		You bet	
		Heads	Tails
Coin falls	Heads	5	−5
	Tails	−5	6

When the problem is presented in this abstract manner, it seems quite obvious that one should bet on tails. This conclusion, however, rests on the assumption that the test-person believes that the coin is as likely to fall heads as tails, and that this belief carries through in his decisions. This is a hypothesis about human behavior which can—and should—be tested experimentally. It is possible that people may have a certain preference for betting "heads," at least in less transparent situations than the one we have considered. Wishful thinking does obviously exist in real life.

Davidson, Suppes, and Siegel test this hypothesis, and find that all their subjects behave rationally in this respect, i.e., that they understand when two events are equally probable, and the implications which this has.

6.9. To design an experiment in any field requires a considerable amount of care to avoid bias of various kinds. When the results of an experiment are published, it is necessary to give a full description of the design, so that a reader can evaluate how significant the results are and if desirable repeat the experiment. In this respect, the book by Davidson, Suppes, and Siegel is a model of precision, even if the particular design they chose may be debatable.

The authors obtained their subjects through the Stanford University Student Employment Service, among students who stated that they were willing to do "relatively unattractive" jobs, such as stapling documents. When a student recruited for such work reported for duty, he was told that the subject for a certain experiment had failed to turn up. He was then asked if he would take part in the experiment instead of stapling documents for two hours at $1 an hour. It was explained to the student that the experiment involved some gambling, and that he might earn less than $2 by two-hour participation. He was, however, assured that the average earning of the participants in the experiment would be more than $2.

The purpose of this cloak-and-dagger tactics was to obtain a "random sample." If the subjects of the experiment had been chosen from a class in "Decision Theory," or recruited by an "honest" advertisement, one would in all probability have obtained a sample which would not have been representative of the student population as a whole. It is, however, not obvious that this would have reduced the value of the experiment. Students from the business school may well have performed differently— better or worse—than students of classics, and this would have been a very interesting observation. The purpose of the whole experiment was presumably to obtain knowledge which has some relevance to economic decisions in real life. It may then be argued that the behavior of a student preparing for a business career is of more interest than the behavior of a "typical" student on the campus.

Of 20 students recruited in this way, 19 agreed to take part in the experiment. The 20th preferred to staple documents for two hours and walk away with his two dollars. He was never heard of again, and it is in some sense to be regretted that he was let off so easily. It would have been interesting to have had him explain his decision.

6.10. We shall follow Davidson, Suppes, and Siegel and use the notation (a, b) for a prospect with a 50–50 chance of getting either a or b.

As bench marks the experimenters took $a = -4$ cents and $b = 6$ cents. They then determined a number c such that the two prospects $(-4, -4)$ and $(6, -c)$ are equivalent, i.e., c is defined by the relation

$$(-4, -4) \sim (6, -c)$$

As a next step they introduced a number d, defined by

$$(6, 6) \sim (-4, d).$$

This was done by asking the subjects to select the most preferred from pairs of bets of the type described in 6.8. The subject had to make a decision; i.e., he was not allowed to state that he was indifferent between two bets.

The bets were presented in an order which did not reveal any systematic pattern.

Since only a finite number of tests could be made, this procedure could only give interval estimates for c and d. If, for instance, the subject decides that

$$(-4, -4) \text{ is preferred to } (6, -15)$$

and

$$(6, -10) \text{ is preferred to } (-4, -4),$$

we can only conclude that there is a number c in the interval $(10 \leq c \leq 15)$ such that

$$(-4, -4) \sim (6, -c).$$

Similarly we can obtain an interval for the value of d which satisfies the condition

$$(6, 6) \sim (-4, d).$$

If a utility function exists, it must satisfy the condition

$$\tfrac{1}{2}u(6) + \tfrac{1}{2}u(6) = \tfrac{1}{2}u(-4) + \tfrac{1}{2}u(d)$$

or

$$u(d) = 2u(6) - u(-4)$$

A utility function is, as we have seen in Chapter III, determined only up to a linear transformation. Hence we can choose two values arbitrarily, for instance our two bench marks. This led Davidson, Suppes, and Siegel to select the utility function which satisfies the conditions

$$u(-4) = -1 \quad \text{and} \quad u(6) = 1$$

as a convenient representation. From this it follows immediately that

$$u(-c) = -3 \quad \text{and} \quad u(d) = 3.$$

When c and d have been determined for a subject, we can determine two other numbers, f and g such that

$$(d, f) \sim (6, -c)$$
$$(-c, g) \sim (-4, d).$$

From the first of these relations we obtain

$$u(d) + u(f) = u(6) + u(-c)$$

or

$$u(f) = u(6) + u(-c) - u(d)$$
$$= 1 - 3 - 3 = -5$$

and similarly $u(g) = 5$.

6.11. Of the 19 subjects who took part in the experiment, 4 made decisions which were inconsistent in the sense that the underlying decision rule, if any, could not be represented by a utility function.

The authors remark ([5] p. 66) that two of these subjects showed a "considerable disinclination" to gamble, and that it really had been a mistake to include them in the experiment. These two students would really have preferred to earn \$2 by stapling documents, and should not have been talked into gambling with the money they probably needed.

The other two were very nervous during the experiment. They made a mess of their decisions, and they themselves seemed quite aware of this.

For the remaining 15 persons, the experiment gave four values of the utility functions in addition to the two values which were chosen arbitrarily. For instance, for subject 1 the experimenters found

$$-18 < f < -15$$
$$-11 < c < -10$$
$$11 < d < 12$$
$$14 < g < 18$$

If we take the middle point of each of the intervals, the utility function of this subject is given by the left part of Table 5. The right part of the table

TABLE 5

Subject 1		Subject 2	
x	$u(x)$	x	$u(x)$
-16.5	-5	-32	-5
-10.5	-3	-11.5	-3
-4	-1	-4	-1
6	1	6	1
11.5	3	15	3
16	5	32.5	5

gives the utility function of subject 2, and it is evident that this function represents a preference ordering quite different from that of subject 1.

6.12. Our short summary cannot do full justice to the well-designed experiment of Davidson, Suppes, and Siegel. We shall, however, not discuss it any further, because this would lead us into a number of problems which are of marginal interest to our subject. It may, however, be useful to refer the reader to a paper by Marschak [6], which discusses some of the principles involved, and to two papers by Becker, De Groot, and Marschak [3] and [4], which give examples of more refined experimental techniques.

In the experiment described in the last paper [4], the subjects were asked to choose one of the following prospects:

A: x_1 or x_4,   each with probability $\frac{1}{2}$
B: x_2 or x_3,   each with probability $\frac{1}{2}$
C: $x_1, x_2, x_3,$ or x_4, each with probability $\frac{1}{4}$

where $x_1 < x_2 < x_3 < x_4$. If the subject chooses C, this implies that

$$\tfrac{1}{4}\{u(x_1) + u(x_2) + u(x_3) + u(x_4)\} > \tfrac{1}{2}\{u(x_1) + u(x_4)\}$$
$$\tfrac{1}{4}\{u(x_1) + u(x_2) + u(x_3) + u(x_4)\} > \tfrac{1}{2}\{u(x_2) + u(x_3)\}.$$

If we add these two inequalities, we obtain a contradiction. Hence a person who makes his decisions in accordance with the Bernoulli Principle should always choose either A or B. The choice of C can be defended only if the subject really considers all three prospects as equivalent.

The experimenters prepared 25 different sets of three such prospects. They then asked 62 students, registered for an introductory course in psychology, to make their choices from each of the 25 sets. It turned out that 60 of the students chose at least one C-prospect. Two selected the C-prospect in as many as 19 of the 25 sets.

One half of the subjects were told that the possible gains (ranging from 0 to 90) represented pennies, and that they would be paid the amounts they won from the prospects they chose. The other half were told that the gains represented dollars and were asked to choose as if there would be a real payoff. Statistical analysis of the choices revealed no significant differences in the behavioral patterns of the two groups.

6.13. Experiments of the type we have discussed have a considerable importance—mainly because they throw light on the basic psychological processes behind economic decisions made under uncertainty. They will, for instance, lead us to doubt the universal validity of the consistency assumptions usually made in economic theory. However, most of this work belongs to experimental psychology rather than to economics. The experiments conducted so far have little direct economic significance, and the authors themselves have never claimed this. During an experiment a student may well behave as if the loss of 20 cents was a minor catastrophe. On the basis of this observed behavior, we can construct a utility function which will represent the preference ordering of the student. It is, however, not very likely that this utility function will make it possible to predict the decisions the same student will make when he buys his lunch after the experiment, or when he goes out for a date in the evening.

We can also construct a utility function based on a student's statements as to how he would decide in situations which could lead to losses of thousands of dollars, which he does not have. It is, however, unlikely that this utility function will give us any useful information about the economic behavior of people with money, or about the behavior of the student in later years when he has some money to lose.

6.14. It is obviously important to obtain knowledge about the shape of the utility functions which govern people's decisions under uncertainty. If experiments can give us little information on this point, it is natural to resort to analysis of the economic behavior we can observe in real life. As a first step we shall quote the conclusions, reached by two of the shrewdest observers of economic activity:

(i) *Adam Smith* states flatly, "The chance of gain is by every man more or less over-valued, and the chance of loss is by most men under-valued, and by scarce any man who is in tolerable health and spirits, valued more than it is worth" ([12] Book I, Chap. X). In Adam Smith's terminology a chance of gain is "worth" its expected value. Hence his statement means, in our terms, that most people have a "risk preference," i.e., that their "attitude to risk" has to be represented by a convex utility function (increasing marginal utility of money). We should note that Adam Smith did not arrive at this conclusion by introspection alone. He thought that he had proved his statement by observing that one could only make a modest profit in insurance, but that it was easy to make a fortune by organizing lotteries. If these observations were correct, the natural explanation is that most people are unwilling to pay much more than the "fair" or "net" premium for insurance cover.

(ii) One hundred years later, *Alfred Marshall* comes to exactly the opposite conclusion. He discusses the "evils of uncertainty" and observes that most people are willing to pay quite handsomely to get rid of these evils ([7] Book V, Chap. VII, and Book VI, Chap. VIII). To prove this statement he refers to insurance companies which have "great expenses in advertising and administration," and still make a good profit. If this observation were correct, there must be a considerable number of people who are willing to pay more than net premium for insurance which gives protection against losses.

Marshall wrote at the height of the Victorian Age, when lotteries no longer were a part of respectable economics. His conclusion was, therefore, quite naturally that people had a "risk aversion," i.e., that the utility of money must be represented by a concave function.

6.15. Most modern economists seem to have accepted Marshall's view. They are quite willing to admit that some people like to gamble, so that risk preference undoubtedly exists, but they do not consider this an important element in the economy. The current school of thought is that most respectable people—the people whose opinions matter—have a risk aversion. The evidence one can quote to support this view is quite overwhelming. Casinos may exist, but they are of no real importance in economic life. The economy is essentially made up of "responsible" people who buy insurance and who diversify their investments.

As a counterexample which may throw some doubt on the prevailing view we can refer to the development of *premium bonds* in some European countries (England, Norway, Sweden, etc.). To illustrate the nature of these bonds, let us assume that a government sells one million bonds, each at $100. If the interest rate is 4%, the total annual interest payment will be

$4 million—or $4 on each bond. The idea behind the premium bond is that instead of paying $4 on each bond, a number of bonds, say 400, are drawn at random each year, and each of these receive $10,000. All bonds are eventually reimbursed at par, so the bondholder cannot lose anything except the interest payments.

This means that if a person holds one orthodox bond, maturing in one year, he will receive $104 with certainty at the end of the year. If he holds one premium bond, he has a prospect which will give

either $10,100 with probability 0.0004
or $100 with probability 0.9996

after one year.

A person with risk aversion will obviously prefer the orthodox bond. There is, however, good evidence that many people prefer the premium bond even to the certainty of getting $105. This means that if the current interest on government bonds is 5%, the government can borrow at 4% by issuing premium bonds.

Several governments have made use of such possibilities, without allowing private borrowers to do so—presumably because it is considered slightly immoral to obtain a cheap loan by exploiting people's desire to gamble. There is, however, little doubt that private businessmen are aware of these possibilities, which were seen already by Adam Smith. The corporate bonds with various options and conversion rights attached serve the same purpose as premium bonds, and probably appeal to the same groups of lenders. However, for convertible bonds we must assume that the probability of making a gain is imperfectly known. This presumably makes the purchase of such bonds a "game of skill" rather than a "game of chance," and hence morally acceptable.

6.16. In spite of this counterexample, there can be little doubt that risk aversion dominates in most societies. The development of the modern welfare states can only be explained by some general aversion to risk. The welfare state seeks to guarantee a decent standard of living to everybody, even if through bad luck or handicaps he is unable to contribute much to the National Income. This means, of course, that those who by their luck or skill make a large contribution to the national wealth must share with the less fortunate members of society. If a society, by free choice and democratic procedure, introduces a welfare state, risk aversion must in some way dominate the decision process. There may, however, be a minority of risk lovers who would prefer a society with greater chances and greater risks, and this may explain some of the dissatisfaction with the welfare state which often finds eloquent expression in some of the advanced European countries. Complaints about lack of opportunity are probably inevitable in a society which seeks to provide security for all.

It is generally accepted that insurance will increase "social welfare" if people have aversion to risk. It is obvious, as has been suggested by some writers, that welfare can be further increased by organizing lotteries to accommodate people with risk preference. This may indicate that the perfect welfare state should have a network of casinos where those who want can gamble with their social security checks.

6.17. Observations of actual economic behavior may not always provide reliable information about what the decision-maker really wants. The experiments demonstrate that people can make the wrong decision "by mistake." This may happen if the choice situation is very complex, or if they do not take enough time and care to analyze the situation.

This leads us to distinguish between the two traditional approaches to our problem:

(i) In business administration one tends to take a *normative* attitude, i.e., to look for the best possible decision—the decision which intelligent persons like ourselves would make. We may, however, soon become aware that finding the best decision may involve more work than it is really worth. We may settle for something less, and make the decision which is just good enough, which *satisfies* certain minimum requirements. This means that instead of being *optimizers*, we become *satisfizers*, a term due to Simon [11].

(ii) In general economics we tend to take a more *descriptive* attitude. We want to find out what rules, if any, businessmen follow when they make decisions under uncertainty. If we know these rules, we may be able to *predict* what will happen in the economy as a whole as the collective outcome of the decisions made by a number of individuals, who together make up the economy.

It may be possible to construct a general theory, based on the assumption that businessmen follow crazy decision rules, for instance, that they cut prices only when the moon is full. It is, however, likely that such a theory would not fit facts, i.e., that the observations we can make in the economy contradict some conclusion we can derive from the theory.

When we are building an economic theory, it is simplest to assume that people behave rationally, i.e., that they know their own interests and that the actions we observe are precisely the actions which will advance these interests in the best possible manner.

This may not be a very realistic assumption, but it is not easy to replace it by a better one. To say that people do not always act rationally in business is not a very useful statement. If the statement is to be useful, it must specify when and how, in what circumstances, and how often the rules of rationality are broken. It is likely that knowledge obtained from experiments will eventually turn out to be very useful on this point.

REFERENCES

[1] Allais, M.: "La Psychologie de l'homme rationnel devant le risque," *Journal de la Société de Statistique de Paris*, 1953, pp. 47–72.

[2] Allais, M.: "Le Comportement de l'homme rationnel devant le risque: Critique des postulates et axiomes de l'école americaine," *Econometrica*. 1953, pp. 503–546.

[3] Becker, G. M., M. H. De Groot, and J. Marschak: "Stochastic Models of Choice Behavior," *Behavioral Science*, 1963, pp. 41–55.

[4] Becker, G. M., M. H. De Groot, and J. Marschak: "An Experimental Study of some Stochastic Models for Wagers," *Behavioral Science*, 1963, pp. 199–202.

[5] Davidson, D., P. Suppes, and S. Siegel: *Decision Making—An Experimental Approach*, Stanford University Press, 1957.

[6] Marschak, J.: "Actual versus Consistent Decision Behavior," *Behavioral Science*, 1964, pp. 103–110.

[7] Marshall, A.: *Principles of Economics*, London, 1890.

[8] Mosteller, F. C. and P. Nogee: "An Experimental Measurement of Utility," *Journal of Political Economy*, 1951, pp. 371–404.

[9] Samuelson, P.: "Probability, Utility and the Independence Axiom," *Econometrica*, 1952, pp. 670–678.

[10] Savage, L. J.: *The Foundations of Statistics*, Wiley, 1954.

[11] Simon, H. A.: "*Models of Man—Social and Rational*," Wiley, 1957.

[12] Smith, A.: *An Inquiry into the Nature and Causes of the Wealth of Nations*, Edinburgh, 1776.

Chapter VII

Decisions with Unknown Probabilities

7.1. In the situations we have discussed in the preceding chapters we have assumed that the decision-maker *knew* the probabilities of the different payoffs which could be the result of his action. This assumption made it possible to reduce the decision problem to selecting the "best" from a given set of known probability distributions representing the available prospects.

Such situations may not often occur in real life. A decision-maker who can list all possible outcomes of an action may well feel unable to assign precise probabilities to these outcomes. Many people, practical business-men as well as theoretical economists, feel that this situation is a completely new one, which calls for an approach radically different from the one we have used so far.

The leading exponent of this position is probably Knight [1], who referred to situations of the first kind, i.e., with known probabilities, as decision problems under *risk*. If the probabilities are not known, the situation becomes, in Knight's terminology, a decision problem under *uncertainty*. This distinction between risk and uncertainty has been widely used in economic literature. We have not adopted this convention, and shall not do so. We shall argue that the convention does not serve any useful purpose, either theoretical or practical.

7.2. We shall approach our problem by considering a situation described by the "payoff" matrix in Table 6. At the head of the table we have an

TABLE 6

	E_1	E_2	$\cdots$	E_j	$\cdots$	E_n
a_1	R_{11}	R_{12}	$\cdots$	R_{ij}	$\cdots$	R_{1n}
a_2	R_{21}		$\cdots$		$\cdots$	
$\vdots$						
a_i	R_{i1}		$\cdots$	R_{ij}	$\cdots$	R_{in}
$\vdots$						
a_m	R_{m1}		$\cdots$	R_{mj}	$\cdots$	R_{mn}

exhaustive set E of mutually exclusive events $E_1, \ldots, E_n$, or a list of all possible "states of the world."

In the column to the left we have listed the elements of the set A of possible actions available to the decision-maker. If our decision-maker chooses action $a_i \in A$, and event $E_j \in E$ occurs, his reward or payoff will be R_{ij}.

This model is very general. It is easy to see that a number of the decision problems which we find in real life can be represented by this model—particularly if we allow the sets A and E to become infinite.

Decision problems which in their nature are "sequential" or "dynamic" do not at first sight seem to fit the model. The *strategy* concept of Von Neumann and Morgenstern ([5] p. 79) will, however, make it possible to bring such problems into the framework of the model.

If we assign probabilities $P(E_j)$ to the events E_j, we are obviously back at the problem discussed in Chapter III. In that case we will have to determine the utility function which represents the preferences of the decision-maker and select the action which maximizes expected utility. This means that we have to determine the action a_i which leads to

$$\max_{ai \in A} \left\{ \sum_{j=1}^{n} P(E_j) u(R_{ij}) \right\}.$$

7.3. To bring something new into the model, we must assume that the probabilities $P(E_j)$ are not known, and seek reasonable rules as to how we would decide in such situations. However, a moment's reflection will make it clear that there are some logical difficulties involved in giving meaning to the statement that the probabilities are unknown. If we insist that we are *completely ignorant* as to which of the events $E_1, \ldots, E_n$ will occur, it is hard to escape the conclusion that all the events are equally probable. This again implies that the probabilities become known, and that we have

$$P(E_j) = 1/n \quad \text{for all } j.$$

This argument is usually referred to as the *Principle of Insufficient Reason* and is generally attributed to Laplace. If it is accepted, it will clearly obviate the need for Knight's distinction between risk and uncertainty.

The principle is, however, far from generally accepted. In practice a decision-maker will usually argue that although he does not know the exact probabilities of the different states of the world, he is not so ignorant that he wants to act as if all states were equally probable. To illustrate the point, let us consider the 2×2 matrix

	E_1	E_2
a_1	u_{11}	u_{12}
a_2	u_{21}	u_{22}

where $u_{ij} = u(R_{ij})$.

It may well happen that our decision-maker will state that he knows that state E_1 is more probable than E_2, but he does not know by how much.

We can then write p for the unknown probability of E_1, and introduce a new sequence of states $E'_1, E'_2, \ldots, E'_j, \ldots$, where E'_j is the state $p = p_j$ and

$$\tfrac{1}{2} < p_1 < p_2 < \cdots < p_j < \cdots < 1.$$

The payoff matrix is then

	E'_1	$\cdots$	E'_j	$\cdots$
a_1	$p_1 u_{11} + (1 - p_1)u_{21}$	$\cdots$	$p_j u_{11} + (1 - p_i)u_{21}$	$\cdots$
a_2	$p_1 u_{12} + (1 - p_1)u_{22}$	$\cdots$	$p_j u_{12} + (1 - p_i)u_{22}$	$\cdots$

This gives us a new problem of the same form as the original one. It is, however, not certain that we have by now been able to confuse the decision-maker so much that he accepts the Principle of Insufficient Reason. He may still insist that some values of p_j should be given more weight than others. We can then repeat the operation any number of times in the hope that he will eventually be satisfied. This procedure, which may seem peculiar, will actually lead to a satisfactory solution, as we shall see in Chapter XIV.

7.4. When we try to use the Principle of Insufficient Reason, we may often find it difficult to list the relevant states of the world. If, for instance, two distinct states give the same payoff, no matter what action we take, should they then be considered as different? To illustrate the point, let us consider a banker in New York who is worried about the next British election and its effect on Pound Sterling. Let us assume that his problem can be described by the following payoff matrix, which implies that a tie in the election, with the Liberal party holding the balance, will be as good for Sterling as a Tory victory.

	Labor wins	Tories win	Liberals hold balance
Buy £	Loss	Gain	Gain
Sell £	Gain	Loss	Loss

If this banker knows nothing about English politics, he may assign the same probability, $\tfrac{1}{3}$, to all three states and make his decision on this basis.

The banker may, however, also describe his problem by the payoff matrix

	Labor wins	Labor does not win
Buy £	Loss	Gain
Sell £	Gain	Loss

If he now applies the Principle of Insufficient Reason, he will act as if he knew that the probability of a Labor victory was $\frac{1}{2}$, and not $\frac{1}{3}$ as in the first example.

7.5. It should be clear that our real problem is to formulate and make use of the *partial knowledge* which the decision-maker feels that he has about the probabilities of the different states of the world. As the problem is difficult, we shall bypass it in our first attack, and study decision rules which do not make use of the known or unknown probabilities of the different states of the world.

This gives us a well-defined mathematical problem. Our task is to find a rule for selecting an action a_i as the best, and this rule shall depend only on the elements in the matrix $\{R_{ij}\}$ or on the utility matrix $\{u(R_{ij})\}$. This approach may seem odd, or at least inefficient, to economists, who feel that any knowledge we have, no matter how vague, should be used in rational decision-making. It may, therefore, be useful to recall that this approach to the problem has its origin in the statistical theory for designing and analyzing scientific experiments. Let us, for instance, assume that we make an experiment to find out how often E_1 is likely to occur in the long run. The essential idea is that the experiment must "speak for itself"; i.e., we must be able to reach a conclusion by considering the experimental evidence alone. We may have believed very strongly, before we conducted the experiment, that E_1 would occur in 90% of the observations. This belief should, however, be considered as irrelevant, or "unscientific," and should not influence the conclusion. We must apparently force ourselves to forget what we believe in order to be completely rational or scientific.

7.6. As an example, let us consider a payoff matrix

	E_1	E_2	E_3	E_4
a_1	2	2	0	1
a_2	1	1	1	1
a_3	0	4	0	0
a_4	1	3	0	0

We shall study some decision rules which may occur to a rational person who is contemplating this matrix. We shall assume that he tries to formulate some general principles which can help him in such situations. He may, for instance, want to lay down rules so that future decisions of this kind can be delegated to his assistant, or to his computer.

(i) The first idea of our decision-maker may be to choose the action where the sum of the gains is largest, i.e., to determine

$$\max_i \sum_{j=1}^m R_{ij}.$$

This decision rule will lead him to choose the action a_1. It is easy to see that this rule is equivalent to the Bernoulli Principle if

(a) Preferences are represented by a linear utility function.
(b) All states of the world are considered equally probable.

We shall refer to the rule as the *Laplace rule*, since it is connected with the Principle of Insufficient Reason.

(ii) Let us next assume that our decision-maker is an extreme pessimist who believes that the most unfavorable state of the world is bound to occur. This will lead him to pick out the smallest value in each row, and then select the row (i.e., the action) in which this minimum is greatest; i.e., he will determine

$$\max_i \left\{ \min_j R_{ij} \right\}.$$

In our example this rule will lead to the choice a_2. We shall refer to it as the *NM-rule*, since it plays a fundamental role in game theory, created by Von Neumann and Morgenstern—a theory which we shall discuss in some detail in Chapter IX.

The rule represents extreme pessimism, and it will lead the decision-maker to prefer the row

$$a_2: 1, \ 1, \ 1, \ 1$$

over a row such as

$$a_k: 100, \ 0, \ 50, \ 50.$$

This will be considered as unreasonable by many people, although probably not to an insurance salesman.

(iii) We can, of course, also construct a decision rule for the extreme optimist, which would consist in selecting the row containing the largest element in the matrix. This rule may be considered as unreasonable for the same reasons as the rule discussed above. However, a compromise between the two rules may possibly lead to a "reasonable" decision rule.

Hurwicz has suggested (without actually publishing the suggestion himself) that a decision-maker might represent his "degree of pessimism" by a number α such that $0 \leq \alpha \leq 1$. For each action he might then consider the best and the worst possible outcome, and weight them by α. This means that he would determine

$$\max_i \left\{ \alpha \min_j R_{ij} + (1 - \alpha) \max_j R_{ij} \right\}$$

and choose the corresponding action. If $\alpha = 1$, i.e., if we are 100% pessimists, the *Hurwicz rule* is obviously the NM-rule. If we take $\alpha = \frac{1}{2}$, the Hurwicz rule will, in our example, lead to the choice of a_3.

This rule may not be very attractive from a normative point of view, since it takes into account only the two extreme outcomes for each action. The rule may, however, have some merit in a descriptive theory about how businessmen actually make their decisions. The theory which Shackle [8] has advanced with considerable vigor seems to be based on ideas closely related to those behind the Hurwicz rule.

(iv) Savage [6] has suggested that a decision-maker may seek to minimize the *regret* he will feel if it should turn out that he has made the "wrong" decision.

If, for instance, he chooses a_1, he will feel no regret if event E_1 occurs—he has done as well as he possibly could. If, however, event E_2 should occur, he will regret that he did not choose a_3. The amount of regret which he feels, will be $4 - 2 = 2$.

From the original payoff matrix the decision-maker can then compute the "regret matrix" by subtracting each element from the largest element in its column. This gives the matrix with elements

$$\max_{j} (R_{ij}) - R_{ij},$$

which in our example becomes:

	E_1	E_2	E_3	E_4
a_1	0	2	1	0
a_2	1	3	0	0
a_3	2	0	1	1
a_4	1	1	1	1

He will then look at the rows in this "regret matrix" and pick the row in which the maximal element is the smallest. This leads to the choice of action a_4.

The *Savage rule* may seem odd to an economist, who is used to analyzing problems in terms of profits or gains. It may, however, appear reasonable to a statistician, who naturally thinks of the loss incurred by making the wrong decision. The rule may also have some appeal to a junior executive who is worried about possible blame from his boss.

7.7. The four decision rules we have discussed all make some sense. They can be justified by various theoretical arguments, and it appears quite plausible that intelligent people may lay down and follow some rules of this kind. It is, however, disturbing that the four rules led to four different decisions when applied to our simple example. This means that at least three of the rules must be "wrong," and it should induce us to take a more systematic approach to the problem.

It seems natural to require that a decision rule give a complete preference ordering of the set of available actions, i.e., pick out not only the best, but also the second best, the third best, etc. We can then lay down some general conditions which a reasonable ordering method should satisfy. As we specify more and more conditions, we reduce the set of acceptable decision rules. We may then hope to end up with just one rule as the unique solution to our problem. This approach has been explored in considerable detail by Luce and Raiffa ([2], Chap. 13), who present several different sets of conditions which a good decision rule should satisfy. We shall, however, not discuss this approach any further, mainly because the results of Savage [7], which we shall present later in this chapter, seem to open a more promising approach. In spite of this, it may be useful, before we leave the subject, to give a brief summary of a paper by Milnor [4]. This paper has become a classic, and it represents one of the most elegant applications of abstract mathematics to a deceptively simple problem in the social sciences.

7.8. Milnor's starting point is the $m \times n$ utility matrix, i.e., the matrix with the elements $u_{ij} = u(R_{ij})$. He then lays down the following ten conditions we may want a good decision rule to satisfy:

1. *Ordering.* The rule should give a complete ordering of the actions, i.e., of the elements $a_i \in A$.
2. *Symmetry.* The ordering should be independent of how we number the rows and columns in the matrix.
3. *Strong domination.* a_i is preferred to a_k if u_{ij} is preferred to u_{kj} for all j.
4. *Continuity.* If the matrices $\{u_{ij}^{(r)}\}$ converge to $\{u_{ij}\}$, and $a_i^{(r)}$ is preferred to $a_k^{(r)}$ for all r, the preference is not reversed at the limit.
5. *Linearity.* The ordering should not be changed if the matrix $\{u_{ij}\}$ is replaced by $\{\alpha u_{ij} + \beta\}$, where $\alpha > 0$.
6. *Row Adjunction.* The ordering between old actions should not be changed if we add new actions, i.e., new rows to the matrix.
7. *Column linearity.* The ordering should not be changed if a constant is added to all elements in a column.
8. *Column duplication.* The ordering should not change if a new column, identical with some old column, is adjoined to the matrix.
9. *Convexity.* If a_i and a_k are equivalent, then neither is preferred to an action with payoffs $\frac{1}{2}(u_{ij} + u_{kj})$, $(j = 1, \ldots, m)$.
10. *Special row adjunction.* The ordering between old rows should not change if we add a new row, provided that no component in this row is preferred to the corresponding components of all old rows.

Milnor then proved that none of the four rules which we have discussed satisfy all 10 conditions. This is fairly easy to see:

(i) The *Laplace rule* obviously violates Condition 8, as the banker in 7.4 would discover.

(ii) The *NM-rule* violates Condition 7.

(iii) The *Hurwicz rule*, which is a generalization of the NM-rule, also violates Condition 7. In addition it violates Condition 9.

(iv) The *Savage rule* violates Condition 6.

Next Milnor proved that each of the four rules are characterized by particular subsets of the 10 conditions. The proofs are a bit tricky, so we shall just state the results:

(i) The only rule which satisfies Conditions 1, 2, 3, 6, and 7, is the *Laplace rule*.

(ii) The only rule which satisfies Conditions 1, 2, 3, 4, 6, 8, and 9, is the *NM-rule*.

(iii) The only rule which satisfies Conditions 1, 2, 3, 4, 5, 6, and 8, is the *Hurwicz rule*.

(iv) The only rule which satisfies Conditions 1, 2, 3, 4, 7, 8, 9, and 10, is the *Savage rule*.

7.9. So far our search for good decision rules has been rather frustrating, and it is natural to look for an entirely different approach. Let us, therefore, assume tentatively that our decision-maker states his preferences as he feels them, and acts accordingly, without trying to justify his actions by referring to universally accepted principles. He may simply consider it more important to satisfy himself than to satisfy axioms.

This decision-maker may be perfectly happy following his own preferences until somebody points out that his preference ordering contains some obvious contradictions. To illustrate the kind of contradictions which can occur, we shall discuss a simple example, due to Marschak [3]. Marschak considers a model where there are only three states of the world, and two possible outcomes, D and L, which he dramatically calls Death and Life. A bar of chocolate and nothing would serve equally well as outcomes. In this model there are only $2^3 = 8$ actions which are different, so the situation can be represented by the following payoff matrix:

	E_1	E_2	E_3
a_1	D	D	D
a_2	L	D	D
a_3	D	L	D
a_4	D	D	L
a_5	D	L	L
a_6	L	D	L
a_7	L	L	D
a_8	L	L	L

The available action can be ordered in 8! different ways, and the decision-maker may feel free to take any of these to represent his preference ordering. Some of these orderings are, however, obviously inconsistent.

Let us first assume that our decision-maker states that

$$a_8 \text{ is preferred to } a_1.$$

This implies that he prefers life to death or, on a more trivial plane, that he likes chocolate.

If the decision-maker next states that a_1 is preferred to a_2, we feel that he is inconsistent. Having established the basic fact that he likes chocolate, we expect him to prefer a chance of getting some chocolate to the certainty of getting none. We will also feel that the decision-maker is inconsistent if he prefers a_2 to a_7. The action a_7 offers him a possibility of getting the coveted bar of chocolate which must be at least as good as the possibility offered by a_2.

Let us assume that our decision-maker, having shown a consistent desire for chocolate, states that he prefers a_2 to a_3 and a_3 to a_4. We can then conclude that he believes that a_2 offers him the best chance of getting the chocolate, or that he believes

$$\text{Prob} (E_1) > \text{Prob} (E_2) > \text{Prob} (E_3).$$

This implies that he must also believe

$$\text{Prob} (E_1 U E_2) > \text{Prob} (E_2 U E_3)$$

and hence that he should prefer a_7 to a_5 since a_7 gives the better chance of obtaining the chocolate bar.

If the decision-maker states that he prefers a_5 to a_7, it is clear that he violates some of the rules of the calculus of probability. Another way of expressing this would be to say that there are some inconsistencies in his belief about the probabilities with which the various states of the world might occur.

7.10. From the examples we have discussed, it is clear that a preference ordering can contain

(i) Inconsistencies with respect to preferences over the possible outcomes. This will happen if the ordering contains a sequence such as

$$a_8 \text{ preferred to } a_4 \text{ preferred to } a_5.$$

(ii) Inconsistencies with respect to beliefs about the probabilities of the possible states of the world. This will happen if the ordering contains a sequence such as

$$a_2 \text{ preferred to } a_4 \text{ preferred to } a_5 \text{ preferred to } a_7.$$

It is obvious that there exist orderings which do not contain inconsistencies of any of these two kinds. We can find one such ordering by assigning utilities to all payoffs and probabilities to all states of the world, and then apply the Bernoulli Principle. The important theorem, due to Savage [7], goes the other way.

Savage proved that for all preference orderings over the set of actions which satisfy the consistency conditions, there exist

(i) A real-valued function over the set of outcomes $u(R_{ij})$—unique up to a positive linear transformation.

(ii) A unique function P over the set of events E, with the properties

$$P(E_j) \geq 0 \qquad\qquad E_j \subset E$$

$$P(E_j \cup E_k) = P(E_j) + P(E_k) \quad E_j \cap E_k = 0$$

$$P(E) = P(E_1 \cup E_2 \cup \cdots \cup E_m \cdots) = 1$$

such that a_i is preferred to a_k if and only if

$$\sum_j P(E_j)u(R_{ij}) > \sum_j P(E_j)u(R_{kj}).$$

7.11. We shall not reproduce Savage's proof, because it is lengthy and requires some fairly advanced mathematics. The reason is that the uniqueness part of the theorem can be proved only if the set of events is infinite, and this makes the definition of consistency rather intricate.

The idea behind the proof is quite simple. It is clear that a statement "a_i is preferred to a_k" will give us an inequality

$$\sum_j P(E_j)u_{ij} > \sum_j P(E_j)u_{kj}.$$

A preference ordering will give us a system of such inequalities and equations. This system can be inconsistent, or it can have solutions. If we require

$$\sum_j P(E_j) = 1$$

it is obvious that if u_{ij} is a solution, $Au_{ij} + B$, where $A > 0$, will also be a solution; i.e., the system can only determine u_{ij} up to a positive linear transformation.

A consistent system of inequalities will usually give intervals for the unknowns as solution. If we add independent new inequalities or equations to the system, it may become inconsistent. If this does not happen, we would on intuitive reasons expect the intervals which contain the solution to narrow down, and in the limit to degenerate into points. That this actually happens is just what Savage proves in a rigorous manner.

To illustrate the idea, let us return to the simple example in 7.9 and assume that it has been established that L is preferred to D. From the statement "a_2 is preferred to a_5" it then follows that

$$P(E_1) > P(E_2) + P(E_3)$$
or
$$P(E_1) > \tfrac{1}{2}.$$

If we add the statement "a_3 is equivalent to a_4," it follows that

$$P(E_2) = P(E_3) < \tfrac{1}{4}.$$

In order to make decisions, a person must establish a preference ordering over the sets of actions available to him. Usually it will be possible to "describe" this preference ordering in several formally different ways. The results of Savage imply that the preference ordering is either inconsistent or that it is possible to specify a utility function and a probability distribution, so that the Bernoulli Principle describes the ordering.

REFERENCES

[1] Knight, F. H.: *Risk, Uncertainty and Profit*, Hart, Schraffner, and Marks 1921.
[2] Luce, R. D. and H. Raiffa: *Games and Decisions*, Wiley, 1957.
[3] Marschak, J.: "Scaling of Utilities and Probability," pp. 95–109 in M. Shubik (ed.), *Game Theory and Related Approaches to Social Behavior*, Wiley, 1964.
[4] Milnor, J.: "Games Against Nature," pp. 49–59 in Thrall, Coombs, and Davis (eds.), *Decision Processes*, Wiley, 1954. Reprinted in M. Shubik (ed.), *Game Theory and Related Approaches to Social Behavior*, Wiley, 1964.
[5] Neumann, J. von and O. Morgenstern: *Theory of Games and Economic Behavior*, 2nd ed., Princeton University Press, 1947.
[6] Savage, L. J.: "The Theory of Statistical Decision," *Journal of the American Statistical Association*, 1951, pp. 55–67.
[7] Savage, L. J.: *The Foundations of Statistics*, Wiley, 1954.
[8] Shakle, G. L. S.: *Expectation in Economics*, Cambridge University Press, 1949.

Chapter VIII

Market Equilibrium under Uncertainty

8.1. In the preceding chapters we have discussed the rules which a rational person could be expected to observe when he has to make decisions under uncertainty. If we know these rules, we should be able to predict how a decision-maker will behave in a *given* situation.

In order to predict the actual behavior of a decision-maker, we must know the situation in which he is placed. From the examples we have discussed, it is clear that some of the uncertainty which bedevils the decision problem of one person may well be due to uncertainty about how another person will make his decision. This indicates that in an economic system—or a group of decision-makers—the decision problems of the various individuals may be interrelated in a complicated manner. In order to understand and predict the behavior of the system as a whole, we must study the nature of these interrelations, which in fact constitute the situation which each decision-maker takes as given.

For a general attack on this problem we must specify the information available to each decision-maker, and the possibilities they have for communicating with each other. We shall, however, not discuss these questions in the present chapter. We shall attack the problem with the methods of classical economic theory, and we shall find that these methods, which served well in so many situations, may break down when we bring uncertainty into the model.

8.2. We shall first restate some of the classical results about *Equilibrium of Exchange* in terms suitable for our purpose. These results are due to Walras [6], and can be found in modernized form in most textbooks of economic theory.

We shall consider an economy, or a market, of m persons and n commodities. We shall assume that in the initial situation, person i holds a quantity x_{ij} of commodity j. This means that the *initial situation* of person i is completely described by a vector

$$x^{(i)} = (x_{i1}, \ldots, x_{in}).$$

The initial situation of the market, or the *initial allocation* of commodities, is described by an $m \times n$ matrix

$$X = \begin{bmatrix} x_{11} & \cdots & x_{1n} \\ \vdots & & \vdots \\ x_{m1} & \cdots & x_{mn} \end{bmatrix}$$

Let us now assume that these m persons find it to their advantage to exchange commodities among themselves, and that these transactions lead to a *final allocation* which can be described by a matrix

$$Y = \begin{bmatrix} y_{11} & \cdots & y_{1n} \\ \vdots & & \vdots \\ y_{m1} & \cdots & y_{mn} \end{bmatrix}$$

This means that the *final situation* of person i is described by a vector

$$y^{(i)} = (y_{i1} \quad \cdots \quad y_{in}).$$

The traditional interpretation of this model is that each person brings to the market commodities which he has produced himself and exchanges them against other commodities which he needs. Often one of these commodities is taken as "money." If a person enters the market with only this commodity, he will be a "buyer," and if he comes to the market without money, he will be a "seller."

8.3. Our problem is to determine the final allocation Y when the initial allocation X is given. In order to solve it, classical economic theory introduced a number of assumptions about how the persons *behave* in this situation. These assumptions can be given in many different forms, which are practically equivalent. For our purpose it is convenient to formulate the assumptions as follows:

Assumption 1: Each person has a preference ordering over the set of allocation matrices, and will seek to reach the most preferred of the attainable allocations.

This is a fairly trivial rationality assumption, which really can be reduced to a tautology.

Assumption 2: The preference ordering of person i $(i = 1, 2, \ldots, m)$ can be represented by a utility function $u_i(z_1, \ldots, z_n)$, where $z = (z_1, \ldots, z_n)$ is the commodity vector allocated to person i.

This is not a trivial assumption. It implies that a person, when assigning utility to an allocation, will consider only "his own row" in the allocation matrix. The assumption rules out any dependence between the enjoyment I get out of a commodity vector and the standard of living enjoyed by other members of the community. If we think that efforts "to keep up with the Joneses" play any significant part in the economic situations we want to study, we must reject this assumption. To make things dramatic, we can say that the assumption implies that I get the same enjoyment out of a case of beer, whether my neighbors are starving or swimming in champagne.

The validity of this assumption appears to have been first seriously questioned by Veblen [4]. Recently Galbraith [3] has argued with considerable force that the assumption does not hold in the "affluent society."

Assumption 3: If a vector of *market prices* $p = (p_1, p_2, \ldots, p_n)$ is quoted, each person will take it as given and beyond his control. He will accept that all exchanges of commodities have to take place at these prices.

This is a behavioral assumption, and it is obviously far from trivial. It implies that our persons adjust passively to the given prices. In classical economic theory, the assumption is usually justified by an assumption that the market consists of a large number of small traders.

8.4. The three assumptions mean that person i will seek the solution to the following problem:

Determine the commodity vector $y^{(i)} = (y_{i1}, \ldots, y_{in})$ which maximizes the utility function $u_i(y_{i1}, \ldots, y_{in})$, subject to the condition

$$\sum_{j=1}^{n} p_j y_{ij} = \sum_{j=1}^{n} p_j x_{ij}.$$

This condition is usually referred to as the "budget equation." It states that the *value* of the commodities held by person i will not change when all transactions take place at market prices. The condition is homogeneous in the elements of p. This means that the solution y will not be affected if all prices are changed in the same proportion or, in the usual economic terms, that only *relative prices* matter.

The maximizing problem we have outlined is solved in most textbooks of economic theory. This solution, which will be familiar, is given by the equations

$$\frac{\partial u_i}{\partial y_{ij}} p_1 = \frac{\partial u_i}{\partial y_{i1}} p_j \quad (j = 2, 3, \ldots, n),$$

which we shall write in the abbreviated form

(1) $$p_1 u_{ij} = p_j u_{i1}.$$

The budget equation can be written

(2) $$\sum_{j=1}^{n} p_j(y_{ij} - x_{ij}) = 0.$$

(1) and (2) give us n equations, which in principle can be solved. The solution will give us

$$y^{(i)} = (y_{i1}, \ldots, y_{in})$$

as functions of the price vector $p = (p_1, \ldots, p_n)$ and the initial holding $x^{(i)} = (x_{i1}, \ldots, x_{in})$, i.e.,

$$y_{ij} = f_{ij}(p, x^{(i)}).$$

This solution gives us the *demand-supply* function of person i for commodity j.

If we find $y_{ij} > x_{ij}$, person i will be a buyer of commodity j, and if $y_{ij} < x_{ij}$, he will be a seller when the price vector $p = (p_1, \ldots, p_n)$ is given.

8.5. Let us now assume that this problem has been solved by all our m persons, i.e., that person i will want to buy or sell so that his holding of commodity j becomes

$$y_{ij} = f_{ij}(p, x^{(i)}).$$

We are, however, considering a closed model, where commodities can neither be produced nor disappear. This means that the wishes of all persons can be satisfied only if

$$(3) \qquad \sum_{i=1}^{m} y_{ij} = \sum_{i=1}^{m} x_{ij} \quad \text{for all } j.$$

Formally this means that the column sums of the matrices X and Y must be the same.

These conditions give us n equations for the determination of the prices $p_1, \ldots, p_n$. If our m persons take these prices as given, total supply will be equal to total demand for each of the n commodities. We shall refer to the set of prices determined by these equations as the *equilibrium prices*.

We can summarize our results as follows: From the basic assumptions we derived the three sets of equations (1)–(3):

Set (1) consists of $n - 1$ equations for each of the m persons, i.e., a total of $mn - m$ equations.
Set (2) consists of one equation for each of the m persons.
Set (3) consists of one equation for each of the n commodities.

This gives us a total of $mn + n$ equations, which, however, are not independent. If we sum the equations in set (2) over all i, we obtain

$$\sum_{i=1}^{m} \sum_{j=1}^{n} p_j(y_{ij} - x_{ij}) = 0.$$

The same equation is obtained if we multiply the equations in set (3) by $p_1, \ldots, p_j, \ldots, p_n$ and sum over all j.

This means that our system contains at most $mn + n - 1$ linearly independent equations. Hence we can, under certain conditions, select one price arbitrarily, say $p_1 = 1$, and determine the $n - 1$ remaining prices and the mn quantities y_{ij} in the final allocation.

8.6. The argument we have given is due to Walras [6]. His basic approach was to count the equations and make certain that the number of equations was equal to the number of unknowns to be determined. This was a substantial advance over pre-Walrasian economic theory, but it is hardly satisfactory by modern standards. The mere fact that the number of equations and the number of unknowns are equal does not guarantee that a system of equations has a solution—and certainly not that the solution is unique.

There are further complications. Even if the system of equations has a unique solution, it is not certain this solution has an economic meaning. p_j and y_{ij} represent prices and quantities of goods, and must by their very nature be non-negative.

If our system of equations has a meaningful solution, we say that there exists a *competitive equilibrium* in our model. To ensure this existence, the utility functions and the initial allocation must satisfy some fairly sophisticated conditions. These conditions were first studied by Abraham Wald [5], and later in a more general setting by Arrow and Debreu [2]. We shall not discuss these conditions, but before we leave the subject, it may be useful to refer to an example given by Wald ([5] pp. 389–390).

Wald considers a three-person three-commodity market, with the given elements

(1) The marginal utility matrix:

$$\{u_{ij}\} = \begin{bmatrix} \dfrac{1}{x_1} & \dfrac{b - x_2}{x_2{}^2} & 2\dfrac{c - x_3}{x_3{}^2} \\[2ex] \dfrac{1}{x_1{}^2} & \dfrac{1}{x_2} & 0 \\[2ex] \dfrac{1}{x_1{}^2} & 0 & \dfrac{1}{x_3} \end{bmatrix}$$

(2) The initial allocation matrix:

$$\{x_{ij}\} = \begin{bmatrix} a & 0 & 0 \\ 0 & b & 0 \\ 0 & 0 & c \end{bmatrix}$$

It is easy to show that, with these elements given, our equations can be satisfied only by

$$p_2 = p_3 = 0$$

$$y_{11} = a, \quad y_{12} = b, \quad y_{13} = c.$$

Mathematically this may be acceptable, but it is nonsense as economics. The "solution" implies that persons 2 and 3 should give their initial holdings to person 1 without any compensation.

8.7. If a competitive equilibrium exists, there exists a set of prices which will "clear the market." This is in itself interesting, but it inevitably leads to the question of if, or how, the persons in the market arrive at trading at these equilibrium prices.

The usual textbook approach is to say that the market behaves *as if* there were a "market manager." This manager announces a set of prices and declares himself ready to buy and sell any amount of commodities at these prices. His objective is to clear his books, i.e., to satisfy all orders, without being caught with any inventory of unsold goods. He may reach this objective by trial and error. If he sets the price of commodity j at p_j and it turns out that at this price he cannot buy enough of the commodity to satisfy demand, he will increase the price. This will increase the supply and reduce the demand for commodity j. By this process of trial and error he may eventually reach the equilibrium prices.

A simpler interpretation of the model is the following, which goes back to Adam Smith: Every day our persons—for instance the artisans in town and the farmers in the neighborhood—bring the same amount of goods to the market. If they trade at prices different from the equilibrium prices, some of them will leave the market dissatisfied. If a person is not able to sell his goods at the prevailing market price, he will offer the goods at a lower price on the following day. Similarly, if he is not able to buy what he wants, he will bid a higher price the following day. After some days of disappointment, the trading prices will eventually become equal to the equilibrium prices, and everybody will be happy when he leaves the market.

With the latter interpretation it may be natural to talk about an "invisible hand" which brings order in the market, to the benefit of everybody.

If, however, we feel the need for a fictitious market manager, we can just as well ask for a real *broker* who will help in settling the problem which cannot be satisfactorily solved by the "free play of market forces." The international gold price is supposed to be fixed at a meeting in London, where leading brokers will set a price which will bring their buying and selling orders in balance.

8.8. The static Walras model seems to capture some of the essential features of the real world. It is far from satisfactory, but it is hard to improve the model without going all the way to a complete dynamic theory.

It would, for instance, seem worthwhile to study what would happen if our persons began trading with a set of prices different from the equilibrium prices, and after some trading realized that these prices would not clear the market. In itself this is not a difficult problem, but we cannot analyze it properly unless we make very detailed assumptions about our persons, i.e., about their intelligence, the information available to them, etc., so that we can determine when they will decide to trade at different prices. It is, however, clear that this will lead us back to the original problem, but with a different initial allocation. This means that our attempt to introduce more realism into the model just leads us to solve the same problem several times in succession. In a theoretical study it is, therefore, natural to start with the simplest of all possible assumptions, namely full information and complete rationality. Then it is not too unreasonable to assume that prices will be set at the equilibrium point right from the start.

8.9. The behavioral assumptions behind the competitive equilibrium model may be open to severe criticism. The real merit of the model lies, however, in the fact that it leads to a *Pareto optimal* allocation of commodities.

An allocation Y is said to be Pareto optimal if there exists no other allocation $\bar{Y}$ such that

$$u_i(\bar{y}^{(i)}) \geq u_i(y^{(i)})$$

for all i, with at least one strict inequality. If this condition is satisfied, there is no further reallocation which can increase the utility of *all* the persons, or, in other words, nobody can improve his situation except at the expense of some other person. However, if we assume that no rational person will agree to a reduction of his utility, it follows that no further reallocation will take place. This means that the competitive equilibrium in a sense is stable, because it leads to a Pareto optimal allocation. This result is of fundamental importance in economic theory, and we shall indicate how it can be proved.

Let us consider an allocation described by a matrix $\{y_{ij}\}$ and assume that a small reallocation changes the matrix to $\{y_{ij} + dy_{ij}\}$. This reallocation will give person i the following change in utility

$$dU_i = \sum_{j=1}^{n} u_{ij} \, dy_{ij}.$$

If the allocation $\{y_{ij}\}$ represents a competitive equilibrium, it follows from 8.4 that

$$u_{ij} = p_j u_{i1}$$

if we take $p_1 = 1$. We then find

$$dU_i = u_{i1} \sum_{j=1}^{n} p_j \, dy_{ij}.$$

It is obvious that in a closed market we must have

$$\sum_{i=1}^{m} dy_{ij} = 0 \quad \text{for all } j.$$

If $u_{i1} > 0$, we can divide by u_{i1}, and sum over all i, and find

$$\sum_{i=1}^{m} \frac{dU_i}{u_{i1}} = \sum_{j=1}^{n} p_j \sum_{i=1}^{m} dy_{ij} = 0.$$

It then follows that dU_i cannot be positive for all i; i.e., the reallocation cannot increase the utility of one person without reducing the utility of another.

8.10. The *Equilibrium of Exchange* is the simplest of the models studied by Walras. In his general equilibrium theory, Walras introduces production facilities, defined by production functions or transformation functions

(4) $$f_k(z_1, \ldots, z_n) = 0 \quad k = 1, 2, \ldots, r.$$

If $z_j > 0$, we say that Commodity j is an *output* of this production process; if $z_j < 0$, we say that it is an *input*. The profit of operating the process is then

(5) $$P_k = \sum_{j=1}^{n} p_j z_j.$$

It is then assumed that the manager of this facility determines the input-output vector which will maximize profits. His problem is then to maximize (5), subject to the restraint (4).

This problem has a mathematical structure which is essentially the same as the one discussed in 8.4. The solution, which is familiar, is given by the equations

$$p_j \frac{\partial f_k}{\partial z_i} = p_i \frac{\partial f_k}{\partial z_j}.$$

This is really the equation system (1), with marginal productivity instead of marginal utility. Hence we see that the introduction of production does not change the mathematical nature of the model.

We can then close the model by assuming

(i) All commodities are produced by some process, except the ultimate inputs: labor and natural resources.
(ii) All commodities are either consumed, or used as inputs in a process.
(iii) All consumers obtain their income from the production processes, either as wages or as profits.

This will give us a complete economic theory—of static nature. Walras also introduces capital, and manages to give the model a certain dynamic aspect, without changing its mathematical nature. It is then possible to prove, as we did in the preceding section, that a competitive equilibrium will be a Pareto optimal situation also in the more general model. This led some enthusiasts to claim that free competition would bring about the "best of all possible worlds."

8.11. The results of Walras do not necessarily hold when we introduce uncertainty. To illustrate the nature of the difficulties we will run into, let us consider two persons, and assume that each of them owns a business or a prospect.

Person 1 owns business 1, which will give a profit of either 1 or zero with equal probability.

Person 2 owns business 2, which will give a profit of either 2, with probability $\frac{1}{4}$, or zero with probability $\frac{3}{4}$.

If the utility of money to person 1 is $u_1(x)$, he will, according to the Bernoulli Principle, assign the following utility to business 1, i.e., to his initial holding:

$$U_1(1, 0) = \tfrac{1}{2}u_1(0) + \tfrac{1}{2}u_1(1).$$

If the utility of money to person 2 is $u_2(x)$, the utility which he assigns to business 2 is

$$U_2(0, 1) = \tfrac{3}{4}u_2(0) + \tfrac{1}{4}u_2(2).$$

This is, of course, nothing but a special case of the general model, introduced in 8.2. The initial allocation is

$$X = \begin{bmatrix} 1 & 0 \\ 0 & 1 \end{bmatrix}$$

We are interested in the final allocation

$$Y = \begin{bmatrix} y_{11} & y_{12} \\ y_{21} & y_{22} \end{bmatrix}$$

which the two persons will agree upon.

If the two prospects are stochastically independent, the reallocation will give the two persons the following utilities:

$$U_1(y_{11}, y_{12}) = \tfrac{3}{8}u_1(0) + \tfrac{3}{8}u_1(y_{11}) + \tfrac{1}{8}u_1(2y_{12}) + \tfrac{1}{8}u_1(y_{11} + 2y_{12})$$

$$U_2(y_{21}, y_{22}) = \tfrac{3}{8}u_2(0) + \tfrac{3}{8}u_2(y_{21}) + \tfrac{1}{8}u_2(2y_{22}) + \tfrac{1}{8}u_2(y_{21} + 2y_{22}).$$

8.12. To make our discussion concrete, we shall assume that the utility functions have the following forms:

$$u_1(x) = 800x - 100x^2 - 350$$

$$u_2(x) = 800x - 50x^2 - 350.$$

We then find

$$U_1(1, 0) = U_2(0, 1) = 0.$$

It is easy to verify that there exist reallocations which will increase the utilities of both persons. We find, for instance,

$$U_1(0.6, 0.4) = 4$$

$$U_2(0.4, 0.6) = 22.$$

If our two persons are rational, they will try to agree on a reallocation of this kind. Let us take the approach of classical economic theory and assume that the price of business 1 is 1, and that the price of business 2 is p. By this we mean that a share y of business 1 ($0 \le y \le 1$) can be exchanged for a share py of business 2 ($0 \le py \le 1$).

A transaction of this kind will give person 1 the utility $U_1(1 - y, py)$. If he believes that he can do nothing to change the price, he will take it as given and adjust to it as best as he can. This means that he will seek to determine the value of y which maximizes his utility.

It is easy to see that this value is determined by the equation

$$\frac{dU_1}{dy} = 350p - 300 - 100y(2p^2 - p + 1) = 0$$

or

$$y = y_1 = \frac{3.5p - 3}{2p^2 - p + 1}.$$

This means that he will want to exchange a share y_1 of his own business for a share py_1 of business 2.

Similarly we find that the transaction will give person 2 the utility $U_2(y, 1 - py)$. This utility is maximized if he acquires a share

$$y_2 = \frac{7.5 - 6p}{2p^2 - p + 1}$$

in business 1.

If both persons are to have their wishes satisfied, we must obviously have

$$y_1 = y_2.$$

This gives us the following equation to determine the *equilibrium price*

$$3.5p - 3 = 7.5 - 6p.$$

From this we find

$$p = 1.105, \qquad y_1 = y_2 = 0.371$$

and

$$U_1(0.63, 0.41) = 16$$

$$U_2(0.37, 0.59) = 8.$$

This is a Pareto optimal allocation; i.e., there is no other exchange of shares which will give both persons a higher utility.

8.13. The behavioral assumptions leading to this result may seem very artificial. Let us therefore drop these classical assumptions and consider a general exchange of shares—or common stock. Let us assume that our two persons, after some bargaining, agree on an arrangement (x, y) such that person 1 holds the shares x and y of businesses 1 and 2 respectively. This obviously implies that person 2 holds the shares $1 - x$ and $1 - y$.

It is easy to see that this arrangement will give the two persons the utilities

$$U_1(x, y) = 400(x + y) - 50(x^2 + xy + 2y^2) - 350$$

and

$$U_2(1 - x, 1 - y) = 700 - 325x - 275y - 25(x^2 + xy + 2y^2) - 350.$$

[97]

Table 7 gives the utilities of the two persons for some selected values of x and y.

TABLE 7

$U_1(x, y)$ = upper figure, $\quad U_2(1 - x, 1 - y)$ = lower figure

x	y					
	0	0.2	0.4	0.6	0.8	1.0
0	-350	-274	-206	-146	-94	-50
	350	293	232	167	98	25
0.2	-272	-198	-132	-74	-24	18
	284	226	164	98	28	-46
0.4	-198	-126	-62	-6	42	82
	216	157	94	27	-44	-119
0.6	-128	-58	4	58	104	142
	146	86	22	-44	-118	-194
0.8	-72	6	66	118	162	198
	74	13	-58	-121	-194	-271
1.0	0	66	124	174	216	250
	0	-62	-128	-198	-232	-350

8.14. Once we drop the classical behavioral assumptions, our two persons are, theoretically, free to agree on any arrangement (x, y) which satisfies the conditions

$$0 \leq x \leq 1$$

$$0 \leq y \leq 1.$$

However, if they both behave rationally, they will only consider arrangements which satisfy the two conditions

(i) $$U_1(x, y) > U_1(1, 0)$$

$$U_2(1 - x, 1 - y) > U_2(0, 1)$$

(ii) There exists no arrangement $(\bar{x}, \bar{y})$ such that

$$U_1(\bar{x}, \bar{y}) > U_1(x, y)$$

$$U_2(1 - \bar{x}, 1 - \bar{y}) > U_2(1 - x, 1 - y).$$

The first of these conditions implies *individual rationality*, in the sense that no person will agree to an exchange arrangement, unless he gains something by it. The second condition expresses Pareto optimality, or *collective rationality*; i.e., our persons will not agree on an arrangement (x, y) if there exists another arrangement $(\bar{x}, \bar{y})$ which will give them both a higher utility. This means that the only acceptable arrangements in Table 7 are the ones corresponding to

$$x = 0.8, \quad y = 0.2 \quad \text{and} \quad x = 0.6, \quad y = 0.4.$$

8.15. Let us now consider two arrangements (x, y) and $(x + dx, y + dy)$, where dx and dy are small, and form the total differentials

$$dU_1(x, y) = \frac{\partial U_1(x, y)}{\partial x} dx + \frac{\partial U_1(x, y)}{\partial y} dy$$

$$dU_2(1 - x, 1 - y) = \frac{\partial U_2(1 - x, 1 - y)}{\partial x} dx + \frac{\partial U_2(1 - x, 1 - y)}{\partial y} dy.$$

These total differentials express the change in the utilities of our two persons caused by a switch from arrangement (x, y) to $(x + dx, y + dy)$.

Condition (ii) of the preceding section is obviously satisfied if dU_1 and dU_2 have opposite signs for all values of dx and dy, i.e., if we have

$$dU_1 \, dU_2 \leq 0$$

where the equality sign only holds for $dx = dy = 0$.

To simplify our notation we shall write

$$\frac{\partial U_1(x, y)}{\partial x} = U_{1x}, \quad \frac{\partial U_2(1 - x, 1 - y)}{\partial x} = U_{2x}, \quad \text{etc.,}$$

so that our condition can be written

$$(U_{1x} \, dx + U_{1y} \, dy)(U_{2x} \, dx + U_{2y} \, dy) \leq 0$$

or

$$U_{1x}U_{2x}(dx)^2 + (U_{1x}U_{2y} + U_{1y}U_{2x}) \, dx \, dy + U_{1y}U_{2y}(dy)^2 \leq 0.$$

From the theory of quadratic forms we know that this inequality is satisfied if and only if

$$U_{1x}U_{2x} \leq 0$$

$$U_{1y}U_{2y} \leq 0$$

$$(U_{1x}U_{2y} + U_{1y}U_{2x})^2 - 4U_{1x}U_{2x}U_{1y}U_{2y} \leq 0.$$

The first two conditions are obviously satisfied. The third condition can be written

$$(U_{1x}U_{2y} - U_{1y}U_{2x})^2 \leq 0$$

which is satisfied only if

$$U_{1x}U_{2y} = U_{1y}U_{2x}$$

or

$$\frac{U_{1x}}{U_{2x}} = \frac{U_{1y}}{U_{2y}} = -k$$

where k is an arbitrary positive constant. By letting k vary, we obtain the set of all Pareto optimal arrangements.

8.16. In our numerical example we have

$$U_{1x} = 400 - 100x - 50y$$
$$U_{1y} = 400 - 50x - 200y$$
$$U_{2x} = -325 - 50x - 25y$$
$$U_{2y} = -275 - 25x - 100y.$$

From this we find that the Pareto optimal arrangements are given by

$$x = \frac{48 - 41k}{7(k + 2)} \quad \text{and} \quad y = \frac{16 - 9k}{7(k + 2)}.$$

Some of these arrangements and the corresponding utilities are given in Table 8. The arrangement corresponding to the competitive equilibrium,

TABLE 8
Pareto optimal exchanges of common stock

		Shares held by person 1	
U_1	U_2	Business 1	Business 2
0	26	0.588	0.400
5	21	0.600	0.403
10	15	0.613	0.406
15	9	0.626	0.410
16	8	0.628	0.411
20	3	0.638	0.413
25	−2	0.651	0.416

which we found in 8.12, is obviously included in this table. This particular arrangement appears, however, quite arbitrary. We can very well imagine that two rational persons might bargain their way to some of the other arrangements in the central part of the table.

8.17. The solutions we have found are Pareto optimal in the sense that there are no other exchanges of shares which will give both persons a higher utility. However, our two persons might think of other transactions than exchanging shares—or common stock—in the two businesses.

To make this point clear, let us note that in our model there are only three non-trivial "states of the world":

State 1: Business 1 succeeds and business 2 fails. Total profits = 1.
State 2: Business 1 fails and business 2 succeeds. Total profits = 2.
State 3: Both businesses succeed. Total profits = 3.

It is then evident that the real problem of our two persons is to make some optimal arrangement as to how profits shall be divided in each possible state of the world.

Arrow [1] has shown that this problem can be solved within the framework of the classical theory. He did this by taking as the basic "commodity" one monetary unit, payable if and only if a particular state of the world should occur. This gives us a classical two-person, three-commodity model, where the commodities are promissory notes or *Arrow-certificates*. The initial allocation in this model is determined by the matrix

$$X = \begin{bmatrix} 1 & 0 & 1 \\ 0 & 2 & 2 \end{bmatrix}$$

The problem then, is to find the final allocation

$$Y = \begin{bmatrix} y_{11} & y_{12} & y_{13} \\ y_{21} & y_{22} & y_{23} \end{bmatrix}$$

By using the same procedure as in 8.12, we find the equilibrium prices

$$p_1 = 1, \quad p_2 = \tfrac{10}{33}, \quad p_3 = \tfrac{9}{33},$$

and the solution

$$y_{11} = 0.644, \quad y_{12} = 0.949, \quad y_{13} = 1.254.$$

The corresponding utilities are

$$U_1 = 18 \quad \text{and} \quad U_2 = 9.$$

This means that both persons can get a higher utility if they drop the restriction that only common stock can be traded. Mathematically this restriction is expressed by the condition

$$y_{11} + y_{12} = y_{13},$$

i.e., the amount which person 1 gets in state 3 must be equal to the sum of the amounts he gets in states 1 and 2.

8.18. It is worth noting that the solution we found in the preceding section is equivalent to the following arrangement:

(i) The two businesses merge and form one corporation.
(ii) The corporation issues preferred stock, which is entitled to a dividend of 1 whenever possible.
(iii) Profits exceeding 1 are paid as dividend on the common stock of the corporation.

If we then allocate to person 1

64.4% of the preferred stock
30.5% of the common stock,

we obtain just the solution of 8.17.

Person 1 has the higher risk aversion, so it is natural that the optimal allocation brought about by the free play of market forces should place most of the preferred stock in his portfolio. This corresponds to the result in 8.12, which gave person 1 most of the least risky asset, i.e., common stock in business 1.

8.19. The behavioral assumption leading to the results in the preceding paragraph may seem artificial. People are used to the idea of trading common stock, and it is not obvious that they will realize that everybody will be better off if they trade in Arrow-certificates. In our simple example the two persons can reach an optimal situation by trading common and preferred stock, but it is not easy to explain how they will arrive at creating just the right kind of preferred stock.

Let us, therefore, take a more general approach. It is clear that the real problem of our two persons is to reach an agreement as to how total profits are to be divided in each of the three states of the world. Let us assume that they somehow have agreed that person 1 is to receive y_1, y_2, or y_3 respectively if states 1, 2, or 3 should occur. This means, of course, that person 2 will receive $1 - y_1$, $2 - y_2$, or $3 - y_3$ respectively.

It is easy to verify that this arrangement will give person 1 the utility

$$U_1 = 300y_1 + 100y_2 + 100y_3 - 12.5(3y_1{}^2 + y_2{}^2 + y_3{}^2) - 350$$

and person 2 the utility

$$U_2 = 700 - 12.5(21y_1 + 6y_2 + 5y_3)$$
$$- 6.25(3y_1{}^2 + y_2{}^2 + y_3{}^2) - 350.$$

In this case an arrangement is defined by a three-element vector (y_1, y_2, y_3). We can then repeat the argument of 8.15, and determine the set of Pareto optimal arrangements. Going through the arithmetic, we find that these arrangements are given by

$$y_1 = 4 - \frac{11k}{1+k}, \quad y_2 = 4 - \frac{10k}{1+k}, \quad y_3 = 4 - \frac{9k}{1+k}.$$

Some of these arrangements and the corresponding utilities are given in Table 9. It is evident that for any arrangement in Table 8, it is possible to find an arrangement in Table 9 (by interpolation if necessary) which will give both persons a higher utility. This shows that both persons stand to gain by dropping the unnecessary *convention* that trading must be restricted to common stock.

TABLE 9
Pareto optimal allocations

		Profits received by person 1		
U_1	U_2	State 1	State 2	State 3
0	28	0.600	0.908	1.217
5	22	0.612	0.920	1.228
10	17	0.624	0.931	1.238
15	11	0.637	0.943	1.249
16	10	0.640	0.945	1.251
18	9	0.644	0.949	1.254
20	6	0.651	0.955	1.260
25	0	0.663	0.966	1.269

8.20. To illustrate how an unnecessary convention can lead to a sub-optimal arrangement, we shall discuss another example, which may be considered a typical bargaining problem.

Let us assume that a contractor is debating whether he should or should not undertake to do a job which will give a gross profit of either 1 or 3, each with a probability of $\frac{1}{2}$. Out of this gross profit he will have to pay an amount w for the labor he must hire for the job. Let us further assume that the contractor's attitude to risk can be represented by the utility function

$$u_1(x) = 8x - x^2.$$

If the contractor accepts the job, he will obtain the utility

$$U_1(w) = \tfrac{1}{2}u_1(1 - w) + \tfrac{1}{2}u_1(3 - w)$$
$$= 15 - (2 + w)^2.$$

If he does not accept the job, his profit and utility will be zero. Hence, he will accept the job if

$$U_1(w) = 15 - (2 + w)^2 > 0$$

or if

$$w < (15)^{\frac{1}{2}} - 2 = 1.87 = \bar{w}.$$

Let us now look at the situation from the point of view of labor. For the sake of simplicity we shall assume that labor is indifferent to risk, i.e., that its utility function is

$$u_2(x) = x.$$

Hence, if the pay is w, labor obtains the utility

$$U_2(w) = \tfrac{1}{2}w + \tfrac{1}{2}w = w.$$

[103]

In this simple situation all possible pay arrangements are Pareto optimal; i.e., one party can increase its utility only at the expense of the other.

Geometrically these arrangements can be represented by the points on the curve

$$U_1 = 15 - (2 + U_2)^2$$

in the positive quadrant in a U_1U_2-plane. The bargaining behavior of labor and the contractor will determine which particular point on the curve the two parties finally agree upon.

8.21. Let us now assume that it occurs to somebody that *profit sharing* may be to the advantage of both parties. They may then consider an arrangement (w_1, w_2), according to which payment to labor will be

$$w_1 \text{ if gross profit is } 1$$
$$w_2 \text{ if gross profit is } 3.$$

This arrangement will give the contractor the utility

$$V_1(w_1, w_2) = \tfrac{1}{2}u_1(1 - w_1) + \tfrac{1}{2}u_1(3 - w_2)$$
$$= 16 - \tfrac{1}{2}(3 + w_1)^2 - \tfrac{1}{2}(1 + w_2)^2.$$

Labor will obtain the utility

$$V_2(w_1, w_2) = \tfrac{1}{2}(w_1 + w_2).$$

From the results in 8.15 it follows that the Pareto optimal arrangements are determined by the conditions

$$\frac{\partial V_1/\partial w_1}{\partial V_2/\partial w_1} = \frac{\partial V_1/\partial w_2}{\partial V_2/\partial w_2} = -k$$

which in this case reduce to

$$3 + w_1 = 1 + w_2 = 2k$$

or

$$w_2 = 2 + w_1.$$

Hence the Pareto optimal arrangements will give the two parties the utilities

$$V_1(w_1, 2 + w_1) = 16 - (3 + w_1)^2$$
$$V_2(w_1, 2 + w_1) = 1 + w_1.$$

By eliminating w_1, we find that these arrangements can be represented by the points on the curve

$$V_1 = 16 - (2 + V_2)^2$$

in the positive quadrant of the V_1V_2-plane. This curve lies "outside" the curve we found in 8.20. This means that for any fixed-wage arrangement we can find a profit-sharing arrangement which will give both parties a higher utility.

8.22. We have so far discussed some extremely simple examples. It is obvious that these can be generalized, and we shall indicate how this can be done. Let us consider a market with n persons and m "commodities" and assume

(i) There are m "prospects." Prospect i will give a monetary payment x_i which is a variate with the distribution $F_i(x_i)$, $(i = 1, 2, \ldots, m)$. For the sake of simplicity we shall assume that these variates are independent, and that
$$F_i(x_i) = 0 \quad \text{for } x_i < 0.$$

(ii) In the initial situation, person j is entitled to a fraction q_{ij} of the payment from prospect i,
$$\sum_{j=1}^{n} q_{ij} = 1 \quad \text{for all } i,$$
$$q_{ij} \geq 0 \quad \text{for all } i \text{ and } j.$$

(iii) Person j has a preference ordering over the set of all prospects, and this ordering can be represented by a Bernoulli utility function $u_j(x)$.

Person j will then assign the following utility to his initial portfolio of common stock:
$$U_j(q_{1j}, \ldots, q_{mj}) = \int_0^\infty \cdots \int_0^\infty u_j\left(\sum_{i=1}^{n} q_{ij}x_i\right) dF_1(x_1) \cdots dF_m(x_m).$$

We have not introduced cash explicitly in the model. We can do this by assuming that one prospect, say prospect m, is degenerate, i.e., that
$$F_m(x_m) = 0 \quad \text{for } x_m < c$$
$$F_m(x_m) = 1 \quad \text{for } c \leq x_m.$$

We can then interpret c as the total amount of cash in the initial situation, and cq_{mj} as the initial cash holding of person j.

8.23. Let us consider a general arrangement which will give person j a payment of
$$y_j(x_1, x_2, \ldots, x_m) \quad (j = 1, 2, \ldots, n)$$
if payments from the m prospects are $x_1, \ldots, x_m$. This arrangement is determined by n functions $y_1, \ldots, y_n$.

Since we work with a closed model, we must have

$$\sum_{j=1}^{n} y_j(x_1, \ldots, x_m) = \sum_{i=1}^{m} x_i.$$

Writing $z = \sum x_i$, we see that there is no loss of generality if we take y_j to be a function of a single variable z; i.e., the problem is to find an arrangement for dividing the total profits z among our n persons.

The arrangement we have described will give person j the following utility:

$$U_j(y) = \int_0^{\infty} u_j(y_j(z)) \, dF(z).$$

Here $F(z)$ is the distribution of the variate z, i.e., the convolution of the distributions $F_1(x_1), \ldots, F_m(x_m)$.

The Pareto optimal arrangements are then given by the n-tuples of functions $y_j(z)$ which maximize

$$\sum_{j=1}^{n} k_j U_j(y) = \int_0^{\infty} \left\{ \sum_{j=1}^{n} k_j u_j(y_j(z)) \right\} dF(z)$$

subject to the condition

$$\sum_{j=1}^{n} y_j(z) = z.$$

Here $k_1, \ldots, k_n$ are arbitrary positive constants.

It is obvious that the solution we seek must maximize the integrand for every value of z. This means that the problem is reduced to finding the vector $y = (y_1, \ldots, y_n)$ which gives

$$\max \sum_{j=1}^{n} k_j u_j(y_j)$$

subject to

(6)
$$\sum_{j=1}^{n} y_j = z.$$

To solve the problem we form the Lagrangian function

$$\sum_{j=1}^{n} k_j u_j(y_j) + \lambda \left(z - \sum_{j=1}^{n} y_j \right).$$

Differentiating we obtain the first-order conditions for a maximum

$$k_j u'_j(y_j) = \lambda,$$

which we shall write in the form

(7)
$$k_j u'_j(y_j) = k_1 u'_1(y_1) \quad (j = 1, 2, \ldots, n).$$

[106]

If the utility functions are of the conventional form, i.e., $u'_j(x) > 0$ and $u''_j(x) < 0$ for all j, conditions (6) and (7) will determine a unique n-tuple $(y_1, \ldots, y_n)$ for given $k_1, \ldots, k_n$. If we let z vary, the conditions will give us an n-tuple of functions of z which represent a Pareto optimal arrangement. We obtain all such arrangements by letting $k_1, \ldots, k_n$ vary.

8.24. As an example, let us take:

$$n = 2, \quad u_1(x) = x^{\frac{1}{2}}, \quad u_2(x) = x^{\frac{3}{4}}.$$

The Pareto optimal payment functions are then given by

$$k_1\{y_1(z)\}^{-\frac{1}{2}} = k_2\{y_2(z)\}^{-\frac{1}{4}}$$

or

$$\{y_1(z)\}^2 - \left(\frac{k_2}{k_1}\right)^4 (z - y_1(z)) = 0.$$

From this we obtain

$$y_1(z) = (h^2 + 4hz)^{\frac{1}{2}} - h$$

$$y_2(z) = z + h - (h^2 + 4hz)^{\frac{1}{2}}$$

where h is an arbitrary non-negative constant.

It is obvious that a Pareto optimal arrangement of this kind cannot be reached by an exchange of common stock or other familiar securities. The arrangement can be reached if our two persons issue Arrow-certificates and trade in these certificates as if they were commodities in the sense of classical economic theory.

If the variate z can take only a finite number of values, i.e., if the number of states of the world is finite, the number of different Arrow-certificates will also be finite. In this case it is not entirely unreasonable to assume that the classical market mechanism may work and lead our persons to a Pareto optimal arrangement.

If, however, z can take an infinity of different values, for instance if it has a continuous distribution, we will get an infinity of different Arrow-certificates, each with its price. This illustrates the point which we have tried to make several times earlier, namely that the introduction of uncertainty implies a step from the finite to the infinite. We can take the step, and still preserve the framework of the classical theory, if we are prepared to assume that *Homo economicus* has unlimited computational facilities so that he can cope with an infinity of prices and markets.

We may make this heroic assumption, but we may also seek other, non-classical, approaches to economic problems. It is the latter point which we shall take up in the following chapters.

[107]

REFERENCES

[1] Arrow, K. J.: "The Role of Securities in the Optimal Allocation of Risk-bearing," *Review of Economic Studies*, 1964, pp. 91–96. (Translation of a paper published in French in 1952.)

[2] Arrow, K. J. and G. Debreu: "Existence of an Equilibrium for a Competitive Economy," *Econometrica*, 1954, pp. 265–290.

[3] Galbraith, J. K.: *The Affluent Society*, Harvard University Press, 1958.

[4] Veblen, T.: *The Theory of the Leisure Class*, 1899.

[5] Wald, A.: "On some Systems of Equations of Mathematical Economics," *Econometrica*, 1951, pp. 368–403. (Translation of a paper originally published in German in 1936.)

[6] Walras, L.: *Elements of Pure Economics*, Richard D. Irwin, 1954. (Translation from French of the book originally published in Lausanne 1874–1877.)

Chapter IX

The Two-person Zero-sum Game

9.1. In the preceding chapters we have made only a few brief references to the theory of games, created by Von Neumann and Morgenstern [12]. This means in a way that we have put the cart before the horse, and it also means that we have failed to give proper credit to a book which has had a profound influence on the social sciences in our time.

We shall not here try to evaluate the impact of game theory on economics and other social sciences. It is, however, appropriate to acknowledge that practically everything which has been said in the preceding chapters is a paraphrasing or elaboration of ideas first formulated and presented in game theory. The same holds for most of the work by modern authors we have quoted. In fact, little of it could have been written before the appearance of the monumental book by Von Neumann and Morgenstern.

9.2. As an introduction to game theory, let us consider the payoff matrix

	E_1	E_2	E_3
a_1	18	2	0
a_2	1	3	10
a_3	5	4	5
a_4	16	3	2

Let us assume, as in Chapter VII, that we have to choose one of the actions a_1, a_2, a_3, or a_4. Let us now interpret the "events" E_1, E_2, and E_3, not as possible states of the world, but as actions to be taken by some other person, whom we shall refer to as the *opponent*.

If we know nothing about the opponent and his motives, we are clearly back at the situation considered in Chapter VII. In order to obtain a new situation, we shall assume that we know that the opponent is an intelligent person, and that his objective is to make our gain as small as possible. This is obviously important knowledge, and we should find some way of using it to our advantage when we make our decision.

Before we attack the problem, we should note that our assumption does not necessarily imply that the opponent is a perverse person who takes pleasure in doing us harm. He may well be a friendly and pleasant person who finds himself in a situation where his interests are directly opposed to ours; i.e., our gain is his loss. The opponent will then in pure self-defense try to make his loss, i.e., our gain, as small as possible.

9.3. Let us now tentatively apply Laplace's Principle of Insufficient Reason; i.e., let us assume that all actions of the opponent are equally probable. Since we do not know what the opponent will do, this may be a reasonable assumption. This approach clearly leads us to choose a_4, which will give us the expected gain

$$\tfrac{1}{3}(16 + 3 + 2) = 7.$$

If the opponent is intelligent, he may well guess that we reason in this way, and he will then choose E_3, and thus reduce our gain to 2.

If we now use our knowledge about the motives and intelligence of the opponent, we conclude that he will actually choose E_3. Our obvious choice is then a_2, which will give us a gain of 10.

However, if the opponent is so clever that he can guess our initial reasoning, he may also be able to guess our second thoughts. This means that he may counter our decision a_2 by choosing E_1 and thus reduce our gain to 1. Our obvious countermove is then to choose a_1, and score a gain of 18. The opponent may, however, out-guess us again and choose E_3, which will reduce our gain to 0. We may then choose a_2 in the hope of securing a gain of 10—and the circle is complete.

9.4. The argument we used in the preceding paragraph led to a dilemma. If we know that the opponent can only think n steps ahead, we may score a solid gain by pushing our own analysis $n + 1$ steps ahead. If, however, the opponent is just as smart as we are, he may also be able to reason $n + 1$ steps ahead and, by doing so, reduce our gain considerably.

There is also a risk that we may overestimate the intelligence of the opponent. If he gives up after the second round, we may lose considerably by assuming that he can follow our reasoning several more steps ahead.

The dilemma clearly calls for a radically new approach. Let us therefore consider the *minimax* rule, which we discussed in Chapter VII (the NM-rule). In a "game against nature" this rule seemed excessively pessimistic. The rule may, however, be appropriate in a game against an opponent who is out to reduce our gain as much as he possibly can.

To apply the rule we have to determine the smallest element in each row of the following payoff matrix.

	E_1	E_2	E_3	Row minimum
a_1	18	2	0	0
a_2	1	3	10	1
a_3	5	4	5	4
a_4	16	3	2	2
Column maximum	18	4	10	4

We then select action a_3, since this gives the largest *minimum gain* of 4. This rule is obviously "safe." The opponent may guess our decision rule, but he can still do nothing to reduce our gain to less than 4.

It may not always be smart to "play safe." Let us, therefore, study the possible reasoning of the opponent. If he uses the minimax rule, he will determine the largest element in each column of the payoff matrix and chose the column which corresponds to our *smallest maximum gain*, i.e., E_2. By doing so, he makes certain that our gain cannot exceed 4.

From these considerations it follows that the minimax rule in some sense leads to an *equilibrium* pair of decisions: (a_3, E_2). If we know that the opponent will use this rule, our best answer is to use the rule ourselves. No other rule can give us a greater gain. In a sense this means that we can be as smart as the opponent, but we cannot be smarter—if he is as smart as possible.

As another example let us consider the payoff matrix

	E_1	E_2	Row minimum
a_1	0	3	0
a_2	1	2	1
Column maximum	1	3	

Here the minimax rule leads to the decisions (a_2, E_1), which also is an equilibrium. Neither we nor the opponent can gain by a unilateral change of decisions.

9.5. The minimax rule worked very well in the two examples we have considered because the payoff matrices had a *saddle-point*, i.e., an element which at the same time was the maximum in its column and the minimum in its row. Some matrices do not have a saddle-point, as we see from the following example:

	E_1	E_2	Row minimum
a_1	2	0	0
a_2	1	3	1
Column maximum	2	3	

In this situation the minimax rule will lead us to choose a_2 and the opponent to choose E_1. This will give us a gain 1. In this case the minimax will not lead to an equilibrium. If we know that the opponent will use the minimax rule, i.e., that he will choose E_1, it is obviously to our advantage to switch from a_2 to a_1 and increase our gain to 2. A clever opponent should, however, be able to deduce that we will reason in this way and thus choose E_2 and reduce our gain to 0.

This argument can obviously be continued, and we will again find ourselves moving in circles as in the example considered in 9.3.

9.6. To find a way out of the dilemma, let us now fall back on the standard mathematical method of giving a name to the thing we don't know. We shall assume that the opponent will choose

$$E_1 \text{ with probability } y$$
$$E_2 \text{ with probability } 1 - y.$$

If we now choose a_1, our *expected* gain will be

$$v(1, y) = 2y + 0(1 - y) = 2y.$$

If we choose a_2, the *expected* gain will be

$$v(0, y) = y + 3(1 - y) = 3 - 2y.$$

This does not seem to help very much. Let us therefore fall back on the oldest of all decision rules and "draw lots," i.e., leave the decision to chance. Let us do this by constructing some random device which will lead us to decision a_1 with probability x, and to a_2 with probability $1 - x$. This rule will give us the expected gain

$$v(x, y) = xv(1, y) + (1 - x)v(0, y)$$

or

$$v(x, y) = 2(2x - 1)y + 3 - 3x.$$

For $x = \frac{1}{2}$ this expression reduces to

$$v(\tfrac{1}{2}, y) = \tfrac{3}{2}.$$

This means that by choosing $x = \frac{1}{2}$, we can make certain that our expected gain is equal to $\frac{3}{2}$, *regardless of what the opponent does*. It seems that the solution to our problem really is to toss a coin; life may be much simpler than we first thought.

9.7. So far we have only shown that we can make certain that our expected gain becomes $\frac{3}{2}$ by choosing $x = \frac{1}{2}$. We may, however, suspect that we have a possibility of doing even better by being really clever.

Tossing a coin is, after all, not the most sophisticated solution to a decision problem.

The expression we found in the preceding section can be rewritten in the following form

$$v(x, y) = (4y - 3)x - 2y + 3.$$

For $y = \frac{3}{4}$ we find

$$v(x, \tfrac{3}{4}) = \tfrac{3}{2}.$$

This means that by choosing $y = \frac{3}{4}$, the opponent can make certain that our expected gain is $\frac{3}{2}$, *regardless of what we do*. This is, however, exactly the expected gain which we could secure for ourselves by choosing $x = \frac{1}{2}$, i.e., if we make our decision by tossing a coin. This shatters any hope we may have had of doing better by being clever. If the opponent is intelligent, he will see to it that our expected gain does not exceed $\frac{3}{2}$.

9.8. Let us now look back. In 9.6. we began by studying a decision problem with unknown probabilities, similar to the problems discussed in Chapter VII. However, we knew the following:

(i) That the opponent was intelligent.
(ii) The objectives of the opponent.

We then found that by making proper use of this knowledge, we could predict the opponent's behavior or, in other words, determine the unknown probabilities. This meant that our decision was reduced to the simpler type discussed in Chapter III.

We also found that knowing the probabilities was not very useful. If the opponent chooses E_1 and E_2 with probabilities $\frac{3}{4}$ and $\frac{1}{4}$ respectively, our expected gain is $\frac{3}{2}$ no matter what we do. If, however, we know that the opponent will use other probabilities, say $(\frac{1}{2}, \frac{1}{2})$, i.e., that he makes a "mistake," we have some useful knowledge which we can exploit. It is easy to see that

$$v(x, \tfrac{1}{2}) = 2 - x.$$

Hence we can obtain an expected gain of 2 by choosing $x = 0$, i.e., by choosing action a_2.

It is worth noting that it is not useful to know only that the opponent is *not* intelligent, i.e., that he will make some mistake. To make profitable use of such knowledge, we must know something more precise about the kind of mistakes he will make.

9.9. Before we try to generalize this result, it may be useful to discuss a simple economic application. Let us assume that our firm and our only competitor in the market have the choice of either television or newspaper advertising. Let us further assume that our share of the market under the various advertising combinations is given by the payoff matrix

		They	
		N	TV
We	N	60%	40%
	TV	50%	70%

We can interpret the matrix to mean that our firm has the best advertising people. If they meet the competitor "head on," they will always get more than half the market for our firm. Our TV man is particularly good. He can secure us at least half the market no matter what the competitor does.

Since we are considering *expected* gains, it is clear that the strategic aspects of the situation will remain unchanged if we divide all elements in the matrix by 10 and subtract 4. This will give us the matrix

		They	
		N	TV
We	N	2	0
	TV	1	3

which is just the matrix which we discussed in 9.5. It then follows that the solution to our problem is to decide on TV or newspaper advertising by tossing a coin. By transferring back to the original units, we see that this "strategy" will give us an *expected* market share of 55%.

9.10. The president of our company may well be taken aback when his marketing expert—with a Ph.D. in game theory—suggests that the decision should be made by tossing a coin. In his search for a better solution, the president may argue that since our TV man is so good the right thing to do must be to give him a free hand with the company's advertising budget. If, however, the competitor can guess that our president will reason in this way, and push his decision through, the competitor will obviously answer with newspaper advertising and bring our market share down to 50%.

If this is pointed out to our president, he may realize that he must keep the competitor guessing. After further thought he may also realize that the only certain way of achieving this is to remain guessing himself, i.e., to let the decision be made by a random device.

More generally we could argue that any scheme which we may think out—no matter how clever—can also be thought out by the competitor, who then will find the countermove which is best from his point of view.

If our president thinks he has found "the obvious" solution, this solution should be obvious also to the intelligent competitor.

9.11. The solution we found to our problem was an *optimal strategy*, described by a probability distribution which in our example was $(\frac{1}{2}, \frac{1}{2})$. If we use this advertising strategy and the competitor uses his optimal strategy $(\frac{3}{4}, \frac{1}{4})$, our expected market share will be 55%.

The market share is, however, a stochastic variable, and we can get a complete description of the situation only by specifying a probability distribution over all possible outcomes, i.e., over all elements in the matrix. It is easy to see that if we and the competitor both use the optimal strategies, our market share will be

$$40\% \text{ with probability } \tfrac{1}{8}$$
$$50\% \text{ with probability } \tfrac{3}{8}$$
$$60\% \text{ with probability } \tfrac{3}{8}$$
$$70\% \text{ with probability } \tfrac{1}{8}$$

This means that the theory we have developed does not enable us to *predict* the outcome of a game. Since randomized decisions enter as an essential part of rational behavior, predictions are possible only in terms of probability statements, no matter how refined we make our analysis. This means that game theory has brought into economics an *uncertainty principle*, similar to the one brought into physics by the quantum theory.

This may be embarrassing to economists who want to predict the outcome of a conflict situation, but it need not embarrass the parties to the conflict. They may, as we have seen in Chapter III, replace the probability distribution by its certainty equivalent. In this example we have tacitly assumed that our utility was linear in terms of market share. This means that the probability distribution above—interpreted as a "prospect" —is equivalent to the certainty of obtaining a market share of 55%.

9.12. In 9.6 we proved a very special case of the so-called *minimax theorem*. We shall now state the theorem in its full generality, without giving a complete proof.

A general two-person zero-sum game is completely described by an $m \times n$ matrix $\{r_{ij}\}$.

The element r_{ij} is the payoff to player 1 if

(i) Player 1 chooses action i, $(i = 1, 2, \ldots, m)$
(ii) Player 2 chooses action j, $(j = 1, 2, \ldots, n)$.

With these choices, payoff to player 2 is by the definition $-r_{ij}$.

In game theory one usually prefers the term "pure strategy" to the term "action," which we have used in the decision problems discussed in the preceding chapters.

Let us now assume that player 1 uses a *mixed strategy*, defined by a vector $x = \{x_1, \ldots, x_m\}$, where the elements satisfy the conditions

$$x_i \geq 0 \qquad \sum_{i=1}^{m} x_i = 1.$$

This means that he decides on a strategy which consists of choosing action i (or pure strategy i) with probability x_i. This means of course that he does not know himself what he will actually do until he has spun his roulette wheel or thrown his dice, and it obviously means that he will keep his opponent guessing.

Let us next assume that player 2 decides to use a mixed strategy, described by the probability vector $y = \{y_1, \ldots, y_n\}$. It is easy to see that the choice of the strategy pair (x, y) will give player 1 the expected payoff

$$v(x, y) = \sum_{i=1}^{m} \sum_{j=1}^{n} r_{ij} x_i y_j.$$

9.13. If now player 1 takes the pessimistic attitude which turned out to be justified in our simple example, he will for each of his own strategies look at the most effective counterchoice and assume that this will be made by his intelligent opponent. This means that for each vector x he will compute

$$\min_{y} \{v(x, y)\}$$

and then pick the x which gives the largest of these minima; i.e., he will compute

$$\max_{x} \left\{ \min_{y} \{v(x, y)\} \right\}.$$

Let us assume that this maximum is attained for the mixed strategy $\bar{x}$. If player 1 uses this "safest of all strategies," he is assured that his expected gain will not be less than a certain value v_1 no matter what the opponent does, i.e.,

$$v(\bar{x}, y) \geq v_1 \quad \text{for all } y.$$

The expected gain of player 2 is $-v(x, y)$. If he reasons in same way as player 1, he will compute

$$\min_{y} \left\{ \max_{x} \{v(x, y)\} \right\}.$$

Let the solution to this problem be $\bar{y}$. This means that by using the mixed strategy $\bar{y}$, player 2 can make sure that his expected gain does not fall under a certain value $-v_2$, or that the gain of player 1 does not exceed v_2, i.e., that

$$v(x, \bar{y}) \leq v_2 \quad \text{for all } x.$$

The minimax theorem states that

$$v_1 = v_2 = v(\bar{x}, \bar{y}).$$

9.14. This theorem was first proved by Von Neumann [11] in 1928 by extensive use of topological methods. Later Von Neumann himself, and others, published simpler proofs. However, the theorem is by no means elementary. In one of his books Bellman presents the minimax theorem in a summary fashion, as we have done in this chapter, and he adds, "This is neither an intuitive, nor a trivial result, but it is true!" ([1] p. 286). We shall take the same attitude. The theorem has far-reaching implications, and we shall be more interested in these than in the mathematical aspects.

The fact that the theorem is deep is brought out eloquently by a discussion initiated by Frechet [7] in *Econometrica* in 1953. It appears from this discussion that the mathematical problem now known as the two-person zero-sum game was first formulated in 1921 by Emile Borel, who used the term "psychological games." Borel found that the minimax theorem was true for symmetric games where the players had as many as five different pure strategies, but he conjectured that the theorem could not be true in general. Borel's intuition led him to believe that it must be an advantage to know the mixed strategy which the opponent would use, even if this was not brought out in simple cases, such as the example we discussed in 9.8.

Intuition tells us—rightly—that it cannot be a disadvantage to know the mixed strategy which the opponent will use, and the example in 9.8 shows that this knowledge can be valuable. The minimax theorem says that there exists one strategy which is so safe that we can reveal it to the opponent without giving him any additional advantage.

9.15. There are some obvious similarities between the two-person zero-sum game and the linear programming problem and its dual. The two problems are in fact equivalent, as Dantzig [3] has proved, and this provides a relatively easy way to the minimax theorem. We shall not reproduce his proof, but we shall try to make his result plausible by reformulating our original problem.

Let us assume that player 1 wants to make sure that his expected gain is at least w, regardless of what player 2 does, i.e., regardless of which pure strategy he chooses. Player 1 will then try to find a mixed strategy, i.e., a vector $x = \{x_1, x_2, \ldots, x_m\}$ with non-negative elements which satisfies the conditions

$$\sum_{i=1}^{m} r_{ij}x_i \geq w \quad \text{for all } j$$

$$\sum_{i=1}^{m} x_i = 1.$$

His real problem is then finding the greatest value of w, say $\bar{w}$, for which such a vector exists.

Player 2 will similarly seek the smallest value of u, say $\bar{u}$, for which there exists a non-negative vector $y = \{y_1, y_2, \ldots, y_n\}$ which satisfies the conditions

$$\sum_{j=1}^{n} r_{ij}y_j \leq u \quad \text{for all } i$$

$$\sum_{j=1}^{n} y_j = 1.$$

9.16. Let us now, for the sake of simplicity, assume that $\bar{w}$ and $\bar{u}$ are both positive and introduce the new variables

$$s_i = x_i/w \quad (i = 1, 2, \ldots, m)$$

and

$$t_j = y_j/u \quad (j = 1, 2, \ldots, n).$$

The problem of player 1 can then be formulated as follows:

$$\text{Minimize: } 1/w = \sum_{i=1}^{m} s_i$$

subject to the conditions

$$\sum_{i=1}^{m} r_{ij}s_i \geq 1 \quad (j = 1, 2, \ldots, n)$$

$$s_i \geq 0 \quad (i = 1, 2, \ldots, m).$$

This is a standard problem in *linear programming*, and it can be solved by a number of different techniques, for instance, by the Simplex method of Dantzig [4].

The problem of player 2 is

$$\text{Maximize: } 1/u = \sum_{j=1}^{m} t_j$$

subject to the conditions

$$\sum_{j=1}^{n} r_{ij}t_j \leq 1 \quad (i = 1, 2, \ldots, m)$$

$$t_j \geq 0 \quad (j = 1, 2, \ldots, n).$$

This is also a problem in linear programming, and it is the *dual* to the problem of player 1.

The central theorem of linear programming states that if the two problems have solutions, say $\bar{w}$ and $\bar{u}$, then $\bar{u} = \bar{w}$. It is not difficult to prove this result, and it is easy to see that it gives us the minimax theorem for non-generate cases. However, a linear programming problem does not always have a solution; for instance, the conditions which consist of linear inequalities may be inconsistent. The real difficulty lies in proving that the linear programming problems resulting from a two-person zero-sum games always have solutions. These problems are discussed in most advanced textbooks on linear programming, e.g., in the book by Dantzig [5], so we shall not pursue the subject any further.

9.17. In real life there are certainly many situations which can be formulated and analyzed as a two-person zero-sum game. The delightful book by Williams [13] provides a number of striking examples. It is, however, clear that most interesting economic situations cannot be squeezed into this simple pattern. The zero-sum (or constant-sum) condition implies an assumption of interpersonal comparability of utility, which is unrealistic, or at least undesirable, in most economic situations. In the example of 9.9, it is natural to assume that an increase of our market share from 60% to 70% in some sense represents a smaller gain in utility than the corresponding loss to the competitor, who sees his market share reduced from 40% to 30%. This reduction may well spell disaster to the competitor, while our gain may give a very modest profit—after taxes. If we admit such considerations, the solution of the two-person zero-sum game is irrelevant to the problem of our two competing firms.

We should see the solution of the two-person zero-sum game as the formidable advance it is, but we should also realize that it is not more than a stepping stone which may eventually lead to a complete revision of the current economic theory. This is in the spirit of Von Neumann and Morgenstern, who state, "The sound procedure is to obtain first utmost precision and mastery in a limited field, and then to proceed to another somewhat wider one, and so on" ([12] p. 7). In the same context they point out that many economists brush aside the simpler problems—of the type which can be represented as two-person zero-sum games—in order to make statements about the larger and more "burning" economic questions. This impatience with "details" can only delay progress.

9.18. The considerations in the preceding section should lead us to soft-pedal the economic importance of the two-person zero-sum game. It may, however, be useful to show that even this simple model brings to light, and into proper perspective, some elements which many economists have felt must be important, but which they were unable to pin down and bring into a formal analysis. To illustrate this, we shall discuss a simple example, which may have more intuitive appeal than practical significance.

In Chapter VIII we assumed that the utility which a consumer attaches to a given amount of commodities is independent of the commodities available to other consumers. This assumption seems quite reasonable for all commodities which can be considered "necessities," and also for those "luxuries" which enable consumers to lead a richer or more enjoyable life. However, many economists have pointed out that the assumption is unrealistic in general. Such considerations are inherent in Adam Smith's brief reference to the scarcity value of diamonds (*Wealth of Nations*, Book 1, Chap. IV). If diamonds were so plentiful that every woman could have a big collection, the utility of a diamond would probably be the same as that of any other pretty pebble.

9.19. Several economists have tried to formalize these ideas. Keynes notes that the needs of human beings ". . . fall into two classes—those needs which are absolute in the sense that we feel them whatever the situations of our fellow human beings may be, and those which are relative in the sense that we feel them only if their satisfaction lifts us above, makes feel superior to, our fellows" [9]. Keynes seemed to believe that with increasing productivity, the first class of needs could be completely satisfied for the whole world's population, provided that there were no major wars or similar catastrophes. This should imply that the economics of the future will be concerned mainly with needs of the second class. This idea was taken up by Galbraith [8], who concludes that in the "affluent society" the main concern of industry will be to produce goods which satisfy needs belonging to Keynes' second class, i.e., goods which have only a "relative utility." It is quite obvious, as Galbraith so eloquently argues, that in this situation we will need an economic theory, substantially different from the classical theory, developed essentially to analyze "allocation of scarce resources." He also concludes that this new theory must allow for considerably more fluctuation than the relatively stable classical models.

9.20. In order to formalize these ideas mathematically, we shall consider an economy of two persons and two commodities.

We shall assume that person A is the richer, and that he can buy either two units of one commodity, or one unit of each, i.e., that he can buy the commodity vectors (2, 0), (1, 1), or (0, 2). We shall further assume that person B, who is the poorer, has the choice of buying either the commodity vector (1, 0) or (0, 1).

The essential new element which we want to bring into classical model is that the utility which person A attaches to his commodity vector depends also on the vector bought by person B. This condition will be fulfilled if the utility to person A is given by a "payoff" matrix of the type

		B's choice	
		(1, 0)	(0, 1)
	(2, 0)	4	0
A's choice	(1, 1)	3	3
	(0, 2)	0	4

Under the assumptions of classical economic theory, A's utility would be independent of the choice made by B; i.e., the two columns of the matrix would be identical. By assuming that A's utility depends not only on his own choice, but also on the choice of B, we formulate the problem in terms of game theory.

In order to make the game a zero-sum one, let us quite arbitrarily assume that the utility of B for any pair of choices has the same numerical value as the corresponding utility of A, but with opposite sign.

It is easy to see that the optimal strategies are

For A: $\{0, 1, 0\}$
For B: $\{\frac{1}{2}, \frac{1}{2}\}$.

This means that A will buy the commodity vector $(1, 1)$, and B will decide whether to buy $(1, 0)$ or $(0, 1)$ by tossing a coin.

9.21. To give an intuitive interpretation to this utility function, we can assume that commodity 1 is a new car, and commodity 2 a vacation in Europe.

The choices $(2, 0)$ and $(1, 0)$ will then mean that both A and B buy a new car. Since A is the richer of the two, he can buy a flashier and bigger car than B. This will give him a feeling of having "squashed" B, and to this feeling he attaches a high utility, i.e., 4.

Similarly the choices $(0, 2)$ and $(0, 1)$ mean A and B both to go to Europe, and that A's "grand tour" completely eclipses the low-cost vacation taken by B. The payoff matrix tells us that A assigns the utility of 4 also to this event.

If, however, the choices are $(2, 0)$ and $(0, 1)$, B can hold his own. When A demonstrates his shiny new car, B can talk about the art galleries and the exquisite food and women of Paris. It is reasonable to assume that A will assign a low utility, say 0, to this frustrating situation. The choice $(1, 1)$ by A, combined with $(1, 0)$ or $(0, 1)$ by B, means that B matches A on one commodity and simply pretends that he is not interested in the other. This may not be quite convincing, for instance, when both persons drive identical cars and B evidently could not afford a vacation in Europe. It is natural to assume that A will have a pleasant feeling of superiority in this situation, and we will assign it a utility of 3.

9.22. The simple example we discussed above shows that game theory makes it possible to formulate and solve problems which have no place in classical economic theory. Some people may find our example amusing, but they will probably insist that the problem is not really important. Suburban snobbism may well exist, but it is an element which most economists seem to feel can be ignored in a serious theory—at least while the "affluent society" is still some decades away.

The problem may, however, become serious if we look at the situation from the producers' point of view. When our two persons use their optimal strategies, the purchases of person B, and hence total purchases, will be random variables. The producers will then face a situation where total demand can be either of the commodity vectors

$$(2, 1) \quad \text{or} \quad (1, 2),$$

each with probability $\frac{1}{2}$.

This means that the producers will have to make their decisions under uncertainty about what demand will be. This is not a startling observation. Any businessman knows that such uncertainty exists in real life. Economists will usually explain this uncertainty with a reference to the "erratic behavior of the consumer." They may also go one step further and suggest that classical demand theory is unrealistic in assuming rational behavior by the consumer. However, our analysis of the two-person zero-sum game shows that there are situations where it *is* rational to behave in an erratic manner. Hence, as suggested in 9.11, game theory brings an uncertainty principle into the rational economic theory.

The objectives pursued by our consumers may be considered irrational by a philosopher or a moralist. However, if they pursue these possibly irrational objectives in a rational manner, the producer will face a situation which is indeed bewildering, at least if his education stopped with classical economic theory.

9.23. To discuss the problem of the producer, it is desirable to develop a slightly more general model. Let $x = (x_1, \ldots, x_n)$ and $y = (y_1, \ldots, y_n)$ be the commodity vectors bought by our two persons. The idea we want to express, is that the richer person (A) will want these vectors in some sense to be as similar as possible, since this will give a clear demonstration of his superiority. On the other hand, the poorer person (B) will want these vectors to be as dissimilar as possible, so that he can claim that his tastes are different from those of person A, and thus make his inferior standard of living less obvious.

We can formalize this idea by considering a suitable defined *distance* $|x - y|$ between the two vectors. We then assume that person A seeks to minimize this distance and that person B seeks to maximize it. This gives us a game-theoretical problem, and we shall assume that the game has the zero-sum property.

We shall assume that there are only two commodities, with prices p_1 and p_2. If the incomes of our two persons are r and s, we have the two budget equations

$$p_1 x_1 + p_2 x_2 = r$$

$$p_1 y_1 + p_2 y_2 = s,$$

assuming that both spend their whole income on the purchase of the two commodities.

To make our discussion concrete, we shall assume that similarity or distance is defined by

$$U = \frac{1}{2}\left(\frac{p_1 x_1}{r} - \frac{p_1 y_1}{s}\right)^2 + \frac{1}{2}\left(\frac{p_2 x_2}{r} - \frac{p_2 y_2}{s}\right)^2.$$

This assumption is of course quite arbitrary, and its justification is mainly mathematical convenience and its usefulness as an illustration.

Using the two budget equations to eliminate x_2 and y_2, we obtain

$$U = \left(\frac{p_1 x_1}{r} - \frac{p_1 y_1}{s}\right)^2 = (t_1 - t_2)^2.$$

In the last expression t_1 and t_2 are the proportions of their income which the two persons spend on commodity 1.

9.24. In the two-person zero-sum game described in the preceding section

Person A will seek to choose t_1 so that U is minimized.
Person B will seek to choose t_2 so that U is maximized.

In accordance with the usual economic theory we have assumed that x_1 and y_1 and hence t_1 and t_2 are continuous variables. This means that our two persons both have an infinity of pure strategies, and this leads to some mathematical complications.

If the payoff function in such an infinite game is a polynomial, the game is usually referred to as a *polynomial game*. This interesting class of games was first studied by Dresher, Karlin, and Shapley [6]. The particular game in our example has been completely solved by Bohnenblust, Karlin, and Shapley [2], so we shall just sketch the solution.

It is clear that a pair of mixed strategies in our game is given by a pair of probability densities $f(t_1)$ and $g(t_2)$. If these mixed strategies are used, the expected payoff to person A will be

$$W = \int_0^1\int_0^1 (t_1 - t_2)^2 f(t_1)g(t_2)\, dt_1\, dt_2$$

$$= V_1 + V_2 + (E_1 - E_2)^2.$$

Here E_1, E_2, V_1, and V_2 are respectively the means and variances of $f(t_1)$ and $g(t_2)$.

It is obvious that person A, who wants to minimize W, will choose a strategy such that $V_1 = 0$, i.e., a pure strategy, say $t_1 = t$. It also is obvious that person B, who wants to maximize W, will choose V_2 as large as he possibly can without placing any restrictions on his choice of E_2. He can achieve this by using a mixed strategy, consisting of the two extremes, $t_2 = 0$ and $t_2 = 1$, with probabilities $1 - q$ and q. These choices will reduce the expected payoff function to

$$W = q(1 - q) + (t - q)^2 = q + t^2 - 2tq.$$

It is easy to verify that this function has a saddle point for $t = q = \frac{1}{2}$. Hence the optimal strategies are

Person A: Spends equal amounts of his income on the purchase of each of the two commodities.

Person B: Spends his whole income on one commodity, selected by tossing a coin or a similar random device.

If both persons use these optimal strategies, total demand will be either of the two commodity vectors

$$\left(\frac{r}{2p_1}, \frac{r}{2p_2} + \frac{s}{p_2}\right) \quad \text{or} \quad \left(\frac{r}{2p_1} + \frac{s}{p_1}, \frac{r}{2p_2}\right)$$

with equal probability.

9.25. If consumers behave as we have outlined in the preceeding paragraphs, producers will have to live and work with uncertainty. It is then natural to ask what effects this will have on the economy as a whole. It is clear that in order to answer this question we must complete the model and make assumptions about the behavior of the producers.

If the producers are classical economists, who believe in the existence of demand curves, they may try to reduce the price when demand for their product seems to go down.

A less classical minded producer who sees the demand for his product diminish may explain this by changes in taste or fashion. He may know that this is a favorite neoclassical explanation whenever observations do not seem to fit economic theory. The producer may then embark on an advertising campaign to restore demand for his product. He will certainly be encouraged to take this action by advertising agencies, who with some justification claim that they can change tastes and fashions. However, if the apparently diminishing demand is due to the game which consumers are playing against each other, the money spent on advertising will be completely wasted. It would have been better for the producer if he had given the money to his church, or any other body which preaches "Thou shallst not covet the property of thy neighbor."

9.26. In our approach to the problem we shall take the model in 9.23 as our starting point. We shall assume

(i) $r = 2, s = 1$

(ii) Only two prices, $\frac{1}{2}$ and 1 are permitted.

This will give us four possible price combinations, which will lead to the following demand:

Price		Demand
$p_1 = p_2$	$= 1$	(1, 2) or (2, 1)
$p_1 = 1, p_2 = \frac{1}{2}$		(1, 4) or (2, 2)
$p_1 = \frac{1}{2}, p_2 = 1$		(2, 2) or (4, 1)
$p_1 = p_2$	$= \frac{1}{2}$	(2, 4) or (4, 2).

We shall find it convenient to refer to these four price combinations as "states" E_1, E_2, E_3, and E_4. State E_1 corresponds to the simple case discussed in 9.20.

We shall next make the following assumption about the behavior of the producers:

(i) If a producer finds that demand for his product is 1, he will reduce his price in the next period.

(ii) If a producer finds that demand for his product is 4, he will increase his price in the following period.

(iii) If both commodities are demanded in equal quantities, the two prices will be equal in the following period; i.e., either $p_1 = p_2 = 1$ or $p_1 = p_2 = \frac{1}{2}$, with equal probability.

Let us now assume that the economy starts in state E_1 (i.e., $p_1 = p_2 = 1$). Demand will then be either (1, 2) or (2, 1). This means that one of the prices will be reduced in the following period, i.e., that the economy will move either to $E_2(p_1 = 1, p_2 = \frac{1}{2})$ or to $E_3(p_1 = \frac{1}{2}, p_2 = 1)$ with equal probability.

Similarly we see that from E_2 the economy can move to E_3 [if demand is (1, 4)] with probability $\frac{1}{2}$, or [if demand is (2, 2)] to E_1 with probability $\frac{1}{4}$, or to E_4 with probability $\frac{1}{4}$.

9.27. It is easy to see that these fluctuations will go on indefinitely. In general the price movements in our economy can be completely described by a Markov chain (see 2.18) with the transition matrix

$$
\begin{bmatrix}
0 & \frac{1}{2} & \frac{1}{2} & 0 \\
\frac{1}{4} & 0 & \frac{1}{2} & \frac{1}{4} \\
\frac{1}{4} & \frac{1}{2} & 0 & \frac{1}{4} \\
0 & \frac{1}{2} & \frac{1}{2} & 0
\end{bmatrix}
$$

It is also fairly easy to show that the long-run probabilities of the four states are respectively $\frac{1}{6}$, $\frac{1}{3}$, $\frac{1}{3}$, and $\frac{1}{6}$.

If, instead of making assumption (iii) in 9.26, we assume that a producer will leave his price unchanged when demand for his product is 2, we will get an economy where price movements are determined by a Markov chain with the transition matrix

$$\begin{bmatrix} 0 & \frac{1}{2} & \frac{1}{2} & 0 \\ 0 & \frac{1}{2} & \frac{1}{2} & 0 \\ 0 & \frac{1}{2} & \frac{1}{2} & 0 \\ 0 & \frac{1}{2} & \frac{1}{2} & 0 \end{bmatrix}$$

Here E_1 and E_4 are "transient states," and the long-run probabilities are $0, \frac{1}{2}, \frac{1}{2}$, and 0. Once the economy enters E_2 or E_3, it will fluctuate between these two states.

9.28. In these two examples we made fairly conventional assumptions about the behavior of the producers. These assumptions have been criticized as unrealistic by some authors, so it may be worthwhile to consider at least one more unorthodox assumption. Let us assume

Producer 1 increases the price of his product when demand is 4 and cuts the price whenever demand is 1.
Producer 2 "follows the leader" and will always charge the price used by producer 1 in the preceding period.

It is easy to see that this will lead to a Markov chain with transition matrix

$$\begin{bmatrix} \frac{1}{2} & 0 & \frac{1}{2} & 0 \\ \frac{1}{2} & 0 & \frac{1}{2} & 0 \\ 0 & \frac{1}{2} & 0 & \frac{1}{2} \\ 0 & \frac{1}{2} & 0 & \frac{1}{2} \end{bmatrix}$$

Here we find that in the limiting case, all four states have a probability $\frac{1}{4}$.

Another unorthodox, and possibly unrealistic, assumption is that the producers completely understand the nature of the demand for their products. In that case the producers should not get worried over fluctuations in their sales. They can just set prices at the level which, according to their objectives, are optimal, and look calmly at the fluctuating sales.

9.29. The simple examples we have discussed illustrate how game theory makes it possible to give a precise formulation to assumptions which many economists think are essential in a realistic economic theory. One of the most eloquent among these economists is Morgenstern [10], who, more than anybody else, has shown how the unsatisfactory state of affairs could be improved.

Our first step was to explain some fluctuations in demand for consumers' goods. This result is not important in itself, even if our explanation is correct. It is, however, clear that the economy may run into serious trouble if the nature of the fluctuations is not properly understood. If government and private enterprise persist in treating a game-theoretical world as it if it obeyed the laws of classical economic theory, they are likely to get some unpleasant surprises.

If, for instance, the producers we have considered believe in the existence of smooth demand curves, they may easily panic when sales decrease or stay low for some time. If they try to remedy this situation by cutting prices—possibly on the advice of competent economists well versed in classical theory—they may get into trouble. They may also cause serious difficulties in the rest of the economy, for instance, if they make drastic changes in their investment when demand falls or goes up.

The uncertainty principle which game theory has introduced implies that it is not possible to predict demand and prices, and this must necessarily upset some of the habitual thinking in economics. The uncertainty principle implies that once the probability distribution over future states is found, it is not possible to obtain more knowledge and make "better forecasts." This means that economists must be satisfied with a statement such as the following: "The odds are 2 to 1 that a projected steel mill at $100 millions will actually be built." They must realize that a statement of this kind contains all knowledge we can possibly obtain, short of reaching absolute certainty. Nothing is gained, and much is lost, if the statement is reformulated as, "The best estimate of investment in the steel industry is $67 millions." It does not help much if we add that the standard deviation is $47 millions.

If we can appeal to the Law of Large Numbers, and get rid of the uncertainty in this way, we are back in classical economics. Then a statement such as, "It is estimated that 8.2 million automobiles will be sold in the U.S.A. next year" may be useful, and may convey practically all the information we have.

REFERENCES

[1] Bellman, R.: *Dynamic Programming*, Princeton University Press, 1957.
[2] Bohnenblust, H. F., S. Karlin, and L. Shapley: "Games with Continuous, Convex Pay-off," *Annals of Mathematics Studies*, No. 24, Princeton University Press, 1950, pp. 181–192.
[3] Dantzig, G. B.: "A Proof of the Equivalence of the Programming Problem and the Game Problem," pp. 330–335 in T. C. Koopmans (ed.), *Activity Analysis of Production and Allocation*, Wiley, 1951.
[4] Dantzig, G. B.: "Maximization of a Linear Function of Variables Subject to Linear Inequalities," p. 339–347 in T. C. Koopmans (ed.), *Activity Analysis of Production and Allocation*, Wiley, 1951.
[5] Dantzig, G. B.: *Linear Programming and Extensions*, Princeton University Press, 1963.

[6] Dresher, M., S. Karlin, and L. Shapley: "Polynomial Games," *Annals of Mathematics Studies*, No. 24, Princeton University Press, 1950, pp. 161–180.

[7] Frechet, M.: "Emile Borel, Initiator of the Theory of Psychological Games and its Application," *Econometrica*, 1953, pp. 95–96.

[8] Galbraith, J. K.: *The Affluent Society*, Harvard University Press, 1958.

[9] Keynes, J. M.: "Economic Possibilities for our Grand-children," *Essays in Persuasion*, London, 1931.

[10] Morgenstern, O.: "Demand Theory Reconsidered," *Quarterly Journal of Economics*, 1948, pp. 165–201.

[11] Neumann, J. von: "Zur Theorie der Gesellschaftsspiele," *Mathematische Annalen*, 1928, pp. 295–320. English translation: *Annals of Mathematics Studies*, No. 40, Princeton University Press, 1959, pp. 13–42.

[12] Neumann, J. von and O. Morgenstern: *Theory of Games and Economic Behavior*, Princeton University Press, 1944.

[13] Williams, J. D.: *The Compleat Strategyst*, McGraw-Hill, 1954.

Chapter X

The General Two-person Game

10.1. In this chapter we shall drop the zero-sum assumption which we used all through Chapter IX. This generalization means that the game no longer can be described by a single matrix. Instead we need two *payoff functions*:

$$M_1(i, j) = \text{Payoff to player 1}$$

and

$$M_2(i, j) = \text{Payoff to player 2},$$

if

(i) Player 1 chooses pure strategy i, $(i = 1, 2, \ldots, m)$.
(ii) Player 2 chooses pure strategy j, $(j = 1, 2, \ldots, n)$.

If the players use a pair of mixed strategies, determined by the probability vectors $x = \{x_1, \ldots, x_m\}$ and $y = \{y_1, \ldots, y_n\}$, the expected payoff of player 1 will be

$$\overline{M}_1(x, y) = \sum_{i=1}^{m} \sum_{j=1}^{n} M_1(i, j) x_i y_j$$

and that of player 2

$$\overline{M}_2(x, y) = \sum_{i=1}^{m} \sum_{j=1}^{n} M_2(i, j) x_i y_j.$$

The problem of each player is now to find a mixed strategy which maximizes his own expected payoff. In Chapter IX we found that the minimax strategy was the solution to our problem. It is clear, however, that this solution may not be quite satisfactory when the zero-sum condition

$$M_1(i, j) + M_2(i, j) = 0$$

does not hold. The minimax strategy is the decision rule of the extreme pessimist who always looks to the worst possible outcome. This attitude turns out to be justified when the interests of the two players are directly opposed, i.e., when the zero-sum condition holds, but we cannot expect it to work equally well in a more general case.

10.2. As a first illustration of the problems we meet in non-zero-sum games, we shall consider the following simple example:

$$m = n = 2$$

$$M_1(1, 1) = 2, \quad M_1(2, 2) = 1, \quad M_1(1, 2) = M_1(2, 1) = 0$$

$$M_2(1, 1) = 1, \quad M_2(2, 2) = 2, \quad M_2(1, 2) = M_2(2, 1) = 0.$$

It is easy to see that this game can be described by the following payoff table:

		Player 2	
		1	2
Player 1	1	(2, 1)	(0, 0)
	2	(0, 0)	(1, 2)

Let us now consider a pair of mixed strategies $(x, 1 - x)$ and $(y, 1 - y)$; i.e., the two players use their first strategy with probabilities x and y respectively. The expected payoffs are then

$$\overline{M}_1(x, y) = 2xy + (1 - x)(1 - y) = 1 - x - (1 - 3x)y$$
$$\overline{M}_2(x, y) = xy + 2(1 - x)(1 - y) = 2 - 2y - (2 - 3y)x.$$

From these expressions we see that by choosing $x = \frac{1}{3}$, player 1 can secure an expected payoff of $\frac{2}{3}$ for himself regardless of what the opponent does. Player 2 can in the same way make his expected payoff equal to $\frac{2}{3}$ by choosing $y = \frac{2}{3}$.

It is, however, clear that this minimax solution does not have the same stability properties as in the zero-sum game. If player 1 *knows* that player 2 is an incarnate pessimist, bound to use a minimax strategy, he can put his knowledge to profitable use. From the expression

$$\overline{M}_1(x, \tfrac{2}{3}) = \tfrac{1}{3} + x,$$

player 1 will see that by choosing $x = 1$ he can increase his expected payoff to $\frac{4}{3}$. Player 2 may, however, argue in the same way, and by considering

$$\overline{M}_2(\tfrac{1}{3}, y) = \tfrac{4}{3} - y$$

he may conclude that he should choose $y = 0$.

If, however, the two basically pessimistic players both decide to make an attempt at exploiting the opponent's pessimism, they will each get a zero payoff, as

$$\overline{M}_1(1, 0) = \overline{M}_2(1, 0) = 0.$$

This shows that it may be risky to depart from the minimax strategy, even if this strategy seems unduly pessimistic.

10.3. If the two players in our example have some possibility of communicating with each other and coordinating their choices, the game becomes almost trivial. In this case it seems obvious that the players should agree to use either the pair of pure strategies (1, 1) or (2, 2). Both these coordinated choices give a total payoff of 3, and the only problem which remains is to divide this payoff between the players.

These considerations show that the payoff table in 10.2 does not give a complete description of the game. The rules of the game have not been specified, and without these rules we cannot solve our problem, i.e., find the best strategies for the two players. This leads us to distinguish between two classes of games:

(i) A game where the rules are such that each player must choose his strategy—pure or mixed—without any possibility of coordinating his choice with that of the opponent will be called a *non-cooperative game.*

(ii) A game in which the players have some possibilities of coordinating their choices—to their mutual advantage—will be called a *cooperative game.*

This distinction has no significance in a zero-sum game, since the players cannot *both* gain by coordinating their choices.

In cooperative games it is sometimes desirable to specify whether *side-payments* are allowed or not. This distinction will, however, not be very important in the games which we shall discuss in the following.

If in our example side-payments are prohibited, or physically impossible, there will no longer be a total payoff of 3 to be divided between the two players. The players can, however, agree to use either the pure strategy pair $(1, 1)$ or the pair $(2, 2)$, and make the decision by some random device. If the device chooses $(1, 1)$ with probability p and $(2, 2)$ with probability $1 - p$, this will give

Player 1 an expected payoff: $2p + 1 - p = 1 + p$

and

Player 2 an expected payoff: $p + 2(1 - p) = 2 - p.$

Hence there is a total *expected* payoff of 3, which can be divided in different ways by appropriate choice of p.

10.4. The game we have discussed in the two preceding paragraphs is usually known as the "Battle of the Sexes" because of the following interpretation:

"A man, Player 1, and a woman, Player 2, each has two choices for an evening's entertainment. Each can either go to a prize fight or to a ballet. Following the usual cultural stereotype, the man prefers the fight, and the woman the ballet; however, to both it is more important that they go out together than that each see the preferred entertainment" ([6] p. 91).

The game can easily be given economic interpretations. It may, for instance, give a good representation of the following situation:

(i) Two insurance companies are considering advertising campaigns to increase the sale of either medical insurance or ordinary life insurance.

(ii) The market situation is such that advertising will have no effect unless both companies put all their resources into a campaign to promote one of the two kinds of insurance.

(iii) Company 1 will gain most if there is an increase in the demand for medical insurance.

(iv) Company 2 stands to gain most if demand for life insurance should increase.

If such simple, clear-cut situations exist in real life, it seems almost inconceivable that the two parties should decide to play the game in a non-cooperative manner and use their minimax strategies. We would expect the two companies to get together and try to reach agreement on a joint advertising campaign. An experienced businessman is, however, likely to point out that this is only half a solution, and in fact the easier half. The real problem of the two parties is to negotiate or bargain their way to an agreement. This indicates that the theory of cooperative games is to a large extent a theory of *bargaining*, a point which we shall return to in 10.8.

10.5. As another illustration, let us consider two competing firms, and assume the following:

(i) If both firms maintain their selling price, each will make a profit of 1.

(ii) If one firm cuts the price, it will double its profits, provided that the other firm maintains its price. The latter firm will then suffer a loss of 2.

(iii) If both firms cut the price, they will both lose 1.

This is also an almost classical game, which can be given many different interpretations. It is usually known as "The Prisoner's Dilemma" ([6] p. 94). It is easy to see that the situation of the two firms can be represented by the following payoff table:

		Firm 2	
		Maintain price	Cut price
Firm 1	Maintain price	(1, 1)	(−2, 2)
	Cut price	(2, −2)	(−1, −1)

As a cooperative game this situation is trivial. The obvious "solution" is that the firms agree to maintain the price. They may possibly reach this solution by instinct, even if they cannot communicate with each other. If, however, they try a rational analysis, they may come to a different result.

To illustrate this, let us, as in our other examples, assume that the two firms maintain the price with probabilities x and y respectively. This will give the expected payoffs

$$\overline{M}_1(x, y) = -x + 3y - 1$$
$$\overline{M}_2(x, y) = 3x - y - 1.$$

It is easy to see that there is no strategy to secure any player an expected gain which is independent of what the opponent does. It is, however, clear that firm 1, which controls x, must choose $x = 0$ in order to make $\overline{M}_1$ as large as possible, and that firm 2 must choose $y = 0$ in order to maximize $\overline{M}_2$. Hence both firms will cut price and suffer a loss. This is not a very satisfactory outcome, and it is difficult to accept it as the final solution to a game played by two rational persons. It seems tempting to assume that the players will somehow find a way to play the game in a cooperative manner.

10.6. We noted that the Prisoner's Dilemma was in a sense trivial considered as a cooperative game. This may have been a rash statement. To see this, let us assume that the players have agreed that both should use the first strategy, i.e., maintain the price. Then either player can score a gain by breaking the agreement, *provided* that the other player keeps it. If he should suspect the other player of planning to break the agreement, he may, in pure self-defense, break the agreement himself.

If the rules of the game make it impossible for the players to break an agreement once it is concluded, the game may be fairly simple, in the sense that it is reduced to a problem in bargaining. If, however, possibilities of cheating exist, the situation becomes complex. It then appears that the outcome of the game will depend on a number of elements such as the ethical standards of the players, the degree to which they trust each other, the temptation of defecting from an agreement, i.e., the size of the payoff, etc. It is difficult to specify these elements in an operational manner, and it does not seem very promising to continue our general speculation—based on introspection—about how rational people—like us—will behave in such situations.

A more fruitful approach may be to make observations in real life in order to find out how people actually behave in situations which can be considered as two-person non-cooperative games. It may be difficult to find suitable situations of this kind, and this naturally leads us to construct them, i.e., to carry out *controlled experiments*. In such experiments we can observe the behavior of the players in games where rules and payoffs can be varied as we wish.

10.7. We shall discuss experimental games in some detail in Chapter XI when we have developed more of the theory, but it may be useful at the present stage to mention the work by Lave [4] and [5] on games similar to the Prisoner's Dilemma. Lave studied a game with the following payoff table:

		Player 2	
		1	2
Player 1	1	(a, a)	(b, c)
	2	(c, b)	(d, d)

where $c > a > d > b$.

Several sequences of such games were played by groups of under-graduates—the usual subjects of such experiments. Lave found, as was to be expected, that the players tended to play in a cooperative manner, with an increasing tendency to double-crossing in the last plays of a sequence.

In a sequence of n plays, Lave found that the players tended to choose the cooperative decision if

$$k(d - b) < n(a - d).$$

In this inequality k is a parameter which depends on the general attitude of the subjects—or, if we like, on the social pattern of the group of subjects. In his first experiments, with students from Reed College, Lave found $k \cong 3$. In later experiments, with Harvard students, he found a stronger tendency to cooperate, i.e., a lower value of k, regardless of the payoffs. This may indicate a strong belief that one can induce others to responsible behavior only by setting a good example oneself.

Lave's result can hardly be taken as a psychological law, verified by experiments, although his formula has a certain intuitive appeal.

$(a - d)$ represents the gain obtained by cooperation in a single play of the game and $n(a - d)$ the gain from cooperation through the whole sequence of n plays. $(d - b)$ represents the loss incurred if one tries to be cooperative and the opponent does not respond.

It seems reasonable to assume that these two elements will influence the decisions of the players. It may be a little surprising that $(c - a)$, the gain obtained by double-crossing, does not enter Lave's formula. It may be possible to explain this, but we shall not pursue the subject any further.

Experiments of this kind may have a considerable psychological interest, but their significance for economics is questionable. The behavior of a student playing for pennies does not contribute much to our knowledge of economic behavior. It does not help if the experimenter asks the student to behave as if he were, for instance, president of U.S. Steel and had to make decisions involving millions of dollars. This will in Lave's words only give us information about "a subject's individual (almost certainly naive) conception of the way he thinks Roger Blough behaves" ([5] p. 30). To a psychologist this may be interesting, but an economist may well dismiss such information as irrelevant to his problems.

10.8. Non-cooperative games constitute a fascinating subject, which we shall leave with some reluctance. It seems, however, that the cooperative games are the more important ones for economic theory. In our two examples both players stand to gain if they communicate and coordinate their decisions. This means that communication has a value, and it is natural to assume that this value will be realized by the players, unless the situation is such that communications are physically impossible.

To illustrate our discussion in the following, we shall use the game with the payoff table

Player 2

		1	2
Player 1	1	(3, 2)	(0, 0)
	2	(0, 0)	(1, 3)

We first note that

Player 1 can secure an expected gain of $\frac{3}{4}$ for himself by choosing his strategy 1 with probability $\frac{1}{4}$.

Player 2 can secure an expected gain of $\frac{6}{5}$ for himself by choosing his strategy 1 with probability $\frac{3}{5}$.

If side-payments are allowed, it seems obvious that both players should use their first strategies and reach some agreement as to how the total gain of 5 should be divided. This will give the two players the payoffs

(i) $$M'_1 = 5x \quad \text{and} \quad M'_2 = 5(1 - x).$$

Here x is a number which has to be settled by bargaining ($0 \leq x \leq 1$).

It is natural to assume that a rational player will cooperate only if he gains something by cooperation, i.e., gets more than he can secure for himself by using his minimax strategy. This means that x must satisfy the conditions

(ii) $$M'_1 = 5x \geq \frac{3}{4}$$

$$M'_2 = 5(1 - x) \geq \frac{6}{5}.$$

Conditions (i) and (ii) are, of course, of the same nature as those we found in 8.14.

If side-payments are not allowed, two rational players will agree to use the pure strategy pair (1, 1) with probability y and the pair (2, 2) with probability $1 - y$. This will give the payoffs

$$M_1 = 2y + 1 \quad \text{and} \quad M_2 = 3 - y.$$

Here y is a number which must be settled by bargaining, and which must satisfy the conditions

$$2y + 1 \geq \tfrac{3}{4} \quad \text{and} \quad 3 - y \geq \tfrac{6}{5}.$$

These are trivial, since y is a probability.

10.9. Let us now look at the more general problem and assume that the two players meet and discuss how they should coordinate their decisions to their mutual benefit. Each player will then argue for a pair of decisions (pure or mixed strategies) which is favorable to him. He may, of course, use any argument which he believes that the other player will swallow. It seems, however, that in an assembly of rational people, he can hope to achieve something only by advancing arguments of the following two types:

(i) He can threaten to *refuse to cooperate*; i.e., he can threaten to choose his own strategy without any regard to the wishes of the other player and without informing him about his choice.

(ii) He can appeal to some *general principle* of fairness or ethics which he thinks may be acceptable to the other player.

From arguments of the first type it follows that we cannot solve a cooperative game problem without considering the corresponding non-cooperative game. This means that non-cooperative games in some sense are more basic than cooperative games. Any player can refuse to cooperate, and the payoff which he can obtain by this refusal must be an element of his "bargaining power" in the cooperative game.

The most intriguing problem in bargaining theory is to formulate the "general principle" acceptable to rational persons which constitutes the arguments of the second type.

To illustrate the point, we shall return to the example discussed in Chapter VIII. In Table 8 of 8.16 we found the Pareto optimal arrangements which could be reached by an exchange of common stock in two businesses. Ignoring that these arrangements are not really optimal (because of the unnecessary restriction that only common stock can be exchanged), we can seek an assumption which will single out one particular arrangement that two rational persons can be expected to agree upon.

In 8.12 we assumed that the two persons behaved as they should, according to classical economic theory, and we found that this determined a unique arrangement, which gave the two persons the utilities

$$U_1 = 16 \quad \text{and} \quad U_2 = 8.$$

It is, however, possible to introduce a number of alternative assumptions. Let us, for instance, assume that our two persons for some reason have agreed that the two businesses have the same "intrinsic value" (since they offer the same *expected* profit), so that in all fairness the common stock must be traded in the ratio 1:1.

In 8.16 we found that the Pareto optimal arrangements were determined by the relations

$$x = \frac{48 - 41k}{7k + 14} \quad \text{and} \quad y = \frac{16 - 9k}{7k + 14}.$$

Here x and y are the shares which person 1 owns of businesses 1 and 2 respectively. With our new additional assumption, we obviously have $x + y = 1$. This determines a unique, but quite different arrangement, giving the two persons the utilities

$$U_1 = 4 \quad \text{and} \quad U_2 = 22.$$

The corresponding shares are

$$x = 0.598 \quad \text{and} \quad y = 0.402.$$

10.10. The additional assumption which we introduced in the preceding paragraph may seem very artificial. The assumption is, however, equivalent to a principle which is widely applied in reinsurance. When two friendly insurance companies of approximately the same size conclude a reciprocal reinsurance treaty, they will usually agree at the outset that the exchange of portfolios shall be on a "net premium basis." This means that they agree to consider only the *expected* profit on the portfolios which they exchange under the treaty. The implications of an agreement of this kind have been discussed in some detail by Borch [1] and [2].

Our example shows that bargaining over the arrangement which the parties should agree upon can be replaced by bargaining over the general principle which should be applied to settle the conflict situation. It seems that this may often happen in practice. In our example it is quite likely that person 1 will argue for settling the conflict by the rules of free competition and by applying the ideas of classical economic theory, since these will lead to an arrangement favorable to him. Person 2 may, on the other hand, argue in favor of some "fair price" principle, since this will give him a better deal.

In bargaining between labor and management, principles like "equal pay for equal work" or "payment based on productivity" may lead to very different settlements, and either party may argue for the principle which best serves its own interests.

A general principle for resolving conflict situations should be acceptable to all parties. We can formalize this by requiring that the principle should be acceptable as fair to us *before* we know which part we are going to play in the game. Ideas of this kind are behind much of our legislation. We consider a law as "good" if we can accept it no matter how we may come in contact with it: as buyer or seller, as tenant or landlord, as creditor or debtor, etc.

10.11. We shall now discuss a general principle, proposed by Nash [7], which gives a unique solution to the two-person bargaining problem. Nash lays down the following four conditions which an arrangement must satisfy to qualify as a "solution" to the problem:

(i) A solution must be invariant under linear transformations of the utility scale.

(ii) A solution must be Pareto optimal.

(iii) Assume that we have a solution, and reduce the problem by removing some possible arrangements which do not represent a solution. The solution to the original problem must also be a solution to the reduced problem.

(iv) If the problem is completely symmetric, the solution is to divide the gain obtained by cooperation equally between the two parties.

Nash then proves that the only arrangement which satisfies these four conditions is the arrangement which maximizes the *product* of the gains in utility, which the two persons make by cooperating.

Applied to the example from Chapter VIII, the Nash principle gives as solution the exchange of common stock (x, y) which maximizes the product

$$\{U_1(x, y) - U_1(1, 0)\}\{U_2(1 - x, 1 - y) - U_2(0, 1)\}.$$

This solution will give the two persons the utilities

$$U_1 = 8 \quad \text{and} \quad U_2 = 17.$$

This result is different from the two "solutions" which we found in 10.9.

10.12. The Nash conditions can be given with full mathematical precision, and the proof can be carried through in a rigorous manner. We shall not do this, but shall apply it to the example from 10.8, in order to illustrate the essential ideas behind the proof.

In the example from 10.8, the case without side-payments is illustrated by Fig. 9. The possible payoff pairs (M_1, M_2) are represented by the points in the triangle OAB with vertices $(0, 0)$, $(1, 3)$, and $(2, 3)$. The Pareto optimal pairs are represented by the points on the line segment AB. The point C $(\frac{3}{4}, \frac{9}{5})$ represents the payoffs when the game is played in a non-cooperative manner.

Let N be a point on the line AB, and D its projection on the horizontal line through C. We shall choose N so that the product $CD \times DN$ is maximized; i.e., N is the point $(107/40, 173/80)$.

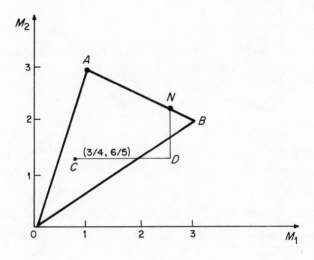

Figure 9

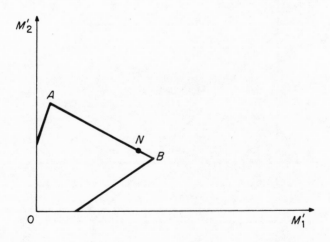

Figure 10

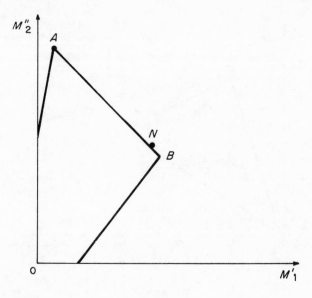

Figure 11

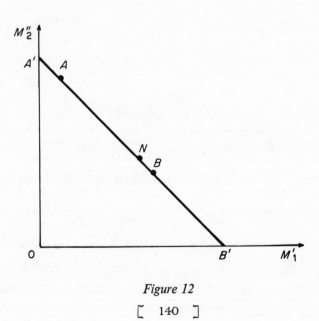

Figure 12

Let us now change the origin on the utility scales so that C becomes the origin; i.e., we introduce new variables defined by

$$M'_1 = M_1 - \tfrac{3}{4}$$

$$M'_2 = M_2 - \tfrac{6}{5}.$$

This will give us the situation, illustrated by Fig. 10. The point N has now become $(77/40, 77/80)$.

Let us next change the utility scale of person 2 by the transformation $M''_2 = 2M'_2$. This will give us the situation, illustrated by Fig. 11. The point N now becomes $(77/40, 77/40)$.

Finally, let us consider an expanded or fictive game where it is possible for the players to reach any payoff pairs in the triangle $OA'B'$ in Fig. 12.

We can now reason backward. From Nash's conditions (ii) and (iv) it follows that the point N represents the unique solution to the symmetric game illustrated by Fig. 12. From condition (iii) it then follows that N also is the solution to the game illustrated by Fig. 11. From condition (i) it further follows that N also represents the solution to the games illustrated by Fig. 10 and Fig. 9.

10.13. The conditions laid down by Nash are simple, and they look very innocent. At first sight it may appear almost self-evident that any arrangement which can be seriously considered as a "solution" must satisfy these four conditions, and one may be a little surprised that there exists only one such arrangement. The Nash conditions may, however, in some cases lead to solutions which many people find unreasonable. As an example, let us consider the game illustrated by Fig. 13. Here the possible arrangements are represented by the points of a triangle with vertices $(0, 0)$, $(1, 1)$, and $(0, 2)$.

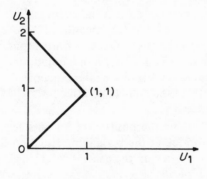

Figure 13

The Nash solution is obviously the point (1, 1). This means, however, that player 1 gets the highest payoff which he can possibly obtain, while player 2 gets only one half of the maximum gain he might have hoped for. If player 2 argues that this is unfair, he implicitly rejects condition (iii), i.e., that the solution shall be independent of "irrelevant alternatives." If player 2 presses his argument, he may threaten to refuse to cooperate altogether and thus reduce the payoff of both players to zero. This means that he is also prepared to reject condition (ii), i.e., that the solution should be a Pareto optimal arrangement.

These considerations indicate that we must take some care in specifying what we really mean by a "solution" to a bargaining situation or, more generally, to a cooperative game.

It may be tempting to require that a solution should enable us to *predict* the actual outcome of a bargaining situation. This is, however, clearly asking too much, unless we accept that predictions can only be made in a probabilistic sense. We found already in our discussion of the two-person zero-sum game that the best prediction we could hope to obtain was a probability distribution over the set of all possible outcomes.

As a predictor the Nash solution is not very attractive, since it considers only Pareto optimal arrangements. There is ample evidence that bargaining situations in real life may lead to non-optimal arrangements. Strikes occur, even if a strike never can be the optimal outcome of a conflict between labor and management.

In our example the player can threaten to refuse cooperation and thus bring about a non-optimal outcome. Such threats cannot be "credible" if at the outset we assume that they will never be carried out in bargaining between rational persons. It seems that if we believe that threats play a real part in bargaining situations, we must accept that the outcome may sometimes be a sub-optimal arrangement.

A solution theory which predicts the outcome of a bargaining situation by specifying a probability distribution over the possible outcomes is a hypothesis which can be tested experimentally. This is a very attractive aspect of any theory, and there exists an extensive literature on so-called experimental games. We have already mentioned the work of Lave [4] and [5]. It may also be useful to refer to an experiment conducted by Stone [9] with students at Stanford University. Stone found that the Nash solution did not give very good predictions of the outcome of a series of games which were essentially of the non-cooperative sort.

10.14. Instead of taking the position of an outsider, trying to predict the outcome, we can look at the bargaining situation from above, as an umpire or arbiter, and draw up the general rules which we will use to resolve the conflict. This leads us to look at the "solution" as an *arbitration scheme*, i.e., a set of rules which will make it possible for the arbiter to propose an arrangement which will be accepted as "fair" by all parties. This idea was first formulated by Raiffa [8], but it is inherent in many earlier papers on game theory. An arbitration scheme will in fact give us just the "general principle" or "additional assumption" which we have been hunting earlier in this chapter.

As an arbitration scheme the Nash solution appears very attractive. It seems almost self-evident that an arbitrated arrangement must be Pareto optimal if it shall be accepted by the players [condition (ii)]. It is also hard to justify an arbitration scheme which does not lead to a symmetric arrangement when applied to a completely symmetric situation [condition (iv)].

The only difficulty occurs in connection with condition (iii), which we now shall express as follows: "If a possible arrangement, which has been rejected, becomes impossible, this should not change the arbiter's decision." This seems a very reasonable condition, but it led to an arrangement, in 10.13, which some people seem to reject as unfair. The following argument may help to clarify this apparent paradox.

If it is possible for a player to inflict a loss on the opponent, he can threaten to do so. Such threat possibilities will strengthen his bargaining position, and this should be reflected in the arbitrated arrangement, which in a sense must include the compensation the player receives for not carrying out his treats. The Nash solution takes this element into account by referring all gains to the minimum payoff which the player can secure for himself regardless of what the opponent does.

On the other hand, it is not obvious that the bargaining position of a player should be weakened if it becomes possible for him to do the opponent a "good turn." Condition (iii) states essentially that such possibilities of exercising charity are irrelevant and should not be taken into account by an arbiter. In practice such possibilities may, however, be considered— on vaguely formulated ethical reasons. It is probably more profitable to sue a "rich" insurance company with a weak case than to sue one's "poor" neighbor with a strong case.

10.15. It is not difficult to find situations in economic life which can be represented as two-person non-zero-sum games. In the following we shall discuss in some detail an example from the distributive trades. The example may have a special interest, since this sector of the economy does not seem to fit very well into the classical economic theory.

We shall begin by considering the problem of a retailer who each Friday has to order goods for resale next week without actually knowing how much he will be able to sell during that week. Usually this retailer will stand to lose if he is caught with an unsold inventory at the end of the week, and he will also lose if he is unable to satisfy the demand from his customers.

We shall assume that he trades in only one commodity, which he orders directly from the producer. Let demand for the commodity be a stochastic variable x with a probability density $f(x) = \alpha e^{-\alpha x}$. Let us further assume that the retailer makes a profit a_1 on each unit he sells of the commodity, and that his storage costs are b_1 per unit. If the retailer holds an inventory y, his expected profit will be:

$$P(y) = \int_0^y a_1 x f(x)\, dx + a_1 y \int_y^\infty f(x)\, dx - b_1 y.$$

Differentiating with respect to y, we find

$$P'(y) = a_1 \int_0^\infty f(x)\, dx - b_1 = a_1[1 - F(y)] - b_1 = a_1 e^{-\alpha y} - b_1.$$

Hence expected profit is maximized if y is determined by the equation:

$$e^{-\alpha y} = b_1/a_1.$$

It is obvious that this solution has a meaning only if $0 \le b_1 \le a_1$.

Assume now that if demand should exceed stocks, the retailer can order an amount up to z from a wholesaler, and that he makes a profit of a_2 on the resale of each of these units. We shall assume that $a_2 < a_1$. We can then interpret $a_1 - a_2$ as the *penalty* our retailer pays per unit of demand which he is unable to satisfy from stock. His expected profit in this situation is

$$M_1(y, z) = a_1 \int_0^y x f(x)\, dx + a_1 y \int_y^{y+z} f(x)\, dx + a_2 \int_y^{y+z} (x - y) f(x)\, dx$$

$$+ (a_1 y + a_2 z) \int_{y+z}^\infty f(x)\, dx - b_1 y$$

$$= \frac{1}{\alpha} \{ a_1(1 - e^{-\alpha y}) + a_2(e^{-\alpha y} - e^{-\alpha(y+z)}) \} - b_1 y.$$

If the retailer knows z, his optimal inventory is determined by the equation

$$\frac{\partial M_1}{\partial y} = a_1 e^{-\alpha y} - a_2 e^{-\alpha y} + a_2 e^{-\alpha(y+z)} - b_1 = 0.$$

10.16. Let us now consider the wholesaler who holds an inventory in the hope of making profits. His sales will be a stochastic variable, which obviously will depend on y, the inventory held by the retailer. If the wholesaler holds an inventory z, his expected profit is

$$M_2(y, z) = a_3 \int_y^{y+z} (x - y) f(x)\, dx + a_3 z \int_{y+z}^\infty f(x)\, dx - b_2 z$$

$$= \frac{1}{\alpha} a_3(e^{-\alpha y} - e^{-\alpha(y+z)}) - b_2 z$$

where a_3 is profit and b_2 storage cost per unit.

If the wholesaler knows y, he will maximize his expected profits by holding an inventory z, determined by the equation

$$\frac{\partial M_2}{\partial z} = a_3 e^{-\alpha(y+z)} - b_2 = 0.$$

In this situation it is clear that the retailer cannot decide on an optimal y unless he knows the value which the wholesaler has chosen for z. Similarly, the wholesaler cannot find an optimal z as long as he does not know the decision made by the retailer. Hence the situation is essentially a two-person game.

Let us now take the following values for the parameters:

$$a_1 = 8, \quad a_2 = a_3 = 4, \quad b_1 = 2, \quad b_2 = 1$$

$$\alpha = \frac{\log 2}{100} = 0.00693.$$

This gives

$$M_1(y, z) = 144\{8 - 4(\tfrac{1}{2})^{y/100} - 4(\tfrac{1}{2})^{(y+z)/100}\} - 2y$$

$$M_2(y, z) = 144\{4(\tfrac{1}{2})^{y/100} - 4(\tfrac{1}{2})^{(y+z)/100}\} - z.$$

In Table 10 $M_1(y, z)$ and $M_2(y, z)$ are calculated for some selected values of y and z. This corresponds to the payoff table in our previous examples.

TABLE 10
Expected profits of the retailer (upper figure)
and of the wholesaler (lower figure)

y	z						
	0	50	100	150	200	300	∞
0	0	170	290	375	430	505	575
	0	120	190	225	230	205	—
50	245	355	440	500	535	595	645
	0	70	100	110	100	55	—
100	375	460	520	560	590	630	665
	0	35	45	35	15	−50	—
150	445	500	545	575	595	620	650
	0	10	0	−20	−40	−120	—
200	465	505	535	555	570	590	610
	0	−10	−30	−60	−90	−175	—
250	450	480	500	515	525	530	550
	0	−20	−50	−80	−125	−210	—

The table illustrates the essential conflict of interest which exists between the two parties. The wholesaler stands to profit if he can induce the retailer to keep a small inventory; the retailer wants the wholesaler to keep the largest possible inventory. The retailer may threaten to increase his inventory and thus reduce the expected profit of the wholesaler. To avoid this the wholesaler may accept to increase his inventory, or he may make the counter-threat of reducing his inventory and thus inflict a loss on the retailer.

10.17. In order to find out how the situation may develop, let us first assume that the retailer believes that for all practical purposes he can act as if $z = \infty$. He may have been led to this belief either by the assurances of the wholesaler, or because the assumption is usually made in the literature on inventory theory. The retailer will then compute

$$\frac{\partial M_1(y, \infty)}{\partial y} = 4(\tfrac{1}{2})^{y/100} - 2$$

and find that his expected profits are maximized for $y = 100$. This is indicated in Table 10.

The more sophisticated wholesaler may know the reasoning of the retailer and hence that his choice will be $y = 100$. The wholesaler will then compute

$$\frac{\partial M_2(100, z)}{\partial z} = 4(\tfrac{1}{2})^{(100 + z)/100} - 1$$

and find that his expected profits are maximized for $z = 100$. If both the retailer and the wholesaler reason in this manner, they will both conclude that 100 is the optimal inventory. If they act accordingly, we shall say that the inventories are in the *state* (100, 100).

If the wholesaler's inventory is $z = 100$, the expected profit of the retailer is 520, and not 665 as he believed when he fixed his own inventory at $y = 100$. However, the retailer may discover this only when demand exceeds 200, i.e., when he has to order more goods than the wholesaler can deliver. The probability that this will happen is

$$\int_{200}^{\infty} \alpha e^{-\alpha x}\, dx = e^{-200\alpha} = (\tfrac{1}{2})^2 = \tfrac{1}{4}.$$

When the retailer discovers that the wholesaler's inventory amounts to only 100, he will increase his own inventory to, say, 150 (see Table 10) in order to maximize expected profits. This means that inventories will move from state (100, 100) to (150, 100).

The wholesaler may in turn discover that the retailer keeps a larger inventory. In order to maximize his profits he must then reduce his own inventory. This means that there will be a transition from state (150, 100) to, say (150, 50). In order to specify the probability of this transition, we must make some additional assumptions about the information available to the wholesaler. We can, for instance, assume that the wholesaler discovers that the retailer has increased his inventory after some time by watching his own order book.

Next time demand exceeds 200, the retailer will discover that the wholesaler has reduced his inventory to 50. We can then repeat our argument, and we see that there will be further transitions to state (200, 50) and finally to state (200, 0), where the process terminates. This means that the wholesaler has been eliminated, and that the retailer has assumed the whole responsibility for keeping inventories.

10.18. To end up in state (200, 0) is obviously not a very satisfactory solution, since there are a number of other states, such as (50, 200), which give both parties a higher expected profit. It seems natural to assume that two rational persons will be able to make a better arrangement. In order to achieve this, they must, however, find some way of coordinating their decisions so that the game can be played in a cooperative manner.

If side-payments are permitted, it is obvious that the first step toward a solution is to maximize total profits, i.e.,

$$M_1(y, z) + M_2(y, z) = 144\{8 - 8(\tfrac{1}{2})^{(y+z)/100}\} - 2y - z.$$

It is easy to see that the maximum is 710, and that it is reached for $y = 0$ and $z = 300$, i.e., in the state (0, 300). This solution means that the two parties agree to make the fullest use of the low storage costs of the wholesaler.

The next step is to agree as to how the expected profits of 710 should be divided between the two parties. We then observe that by keeping an inventory $y = 200$, the retailer can secure an expected profit of 465 for himself regardless of what the wholesaler does.

Similarly the wholesaler can, by choosing $z = 0$, make certain that his expected profit does not become negative. Hence we get the Nash solution by finding

$$\max (M_1 - 465)M_2$$

subject to

$$M_1 + M_2 = 710.$$

The solution is

$$M_1 = 588, \qquad M_2 = 122.$$

If no side-payments are permitted, the Nash solution is given by

$$\max_{y,z} \{M_1(y, z) - 465\}M_2(y, z)$$

subject to $y \geq 0$. The solution is

$$y = 0, \qquad z = 330$$

and

$$M_1 = 515, \qquad M_2 = 185.$$

10.19. So far we have found a number of possible solutions to our problem, and none of them can be considered satisfactory.

In 10.17 we assumed that the two parties behaved as they should according to classical economic theory, i.e., that they analyzed the situation independently, and made the decisions which to the best of their knowledge would give the highest expected profits. This assumption gave us state (100, 100) as the unique solution. We found, however, that this solution was not *stable*. After some periods the two parties would change their decisions, and eventually they would arrive at state (200, 0), and remain there. This state cannot be considered "optimal" in any sense of the word, and it implies that the wholesaler is eliminated. However, wholesalers do exist in a number of trades, so our conclusion is contradicted by observations.

In 10.18 we assumed that the two parties behaved as rational players in a cooperative game. This gave us a solution which implied that the retailer is reduced to an agent, receiving orders which he forwards to the wholesaler. This is also contradicted by observations, since there undoubtedly exist retailers who own inventories and carry the risk involved.

The model we have discussed is extremely simple, and it is not in itself surprising that it gives an inadequate representation of the real world. It seems, however, that a more complicated model will also contain a tendency to move toward extreme positions, where some of the players are eliminated. This point has been studied in another paper [3], and we shall not discuss it any further in the present context. It may, however, be useful to remark that there are no obvious reasons for assuming that it should be optimal to have several independent decision-makers controlling inventories along the channels which bring goods from the producer to the consumer.

10.20. In 8.10 we indicated how *production* could be brought into the classical equilibrium model of Walras. For our present purpose it is convenient to summarize these results as follows: (i) If a set of equilibrium prices is known, and (ii) each consumer buys the commodity vector which will maximize his utility, and (iii) each producer selects the input-output vector which will maximize his profits, then the resulting situation is Pareto optimal.

This means that decision-making can be *decentralized*; i.e., nothing can be gained by coordinating the decisions, and "playing the game" in a cooperative manner.

In the classical theory it is assumed that goods go directly from producers to consumers. Retailers and wholesalers appear as middlemen who have no place in the theory. In real life they do, however, constitute an important sector of the economy, and it is clearly desirable to have a theory which can explain how this sector functions. It seems that this theory would be rather trivial in the absence of any uncertainty, and this is probably why the sector "distributive trades" has been ignored by most economists. If consumers knew their future income with certainty, and made firm and detailed plans as to how they would spend this income, the whole economy could be run as a mail-order business.

When we introduce uncertainty, the function of the distributive trades becomes evident. Goods are usually produced on fairly rigid schedules in modern industry, and even more so in agriculture. If these schedules cannot be synchronized with the needs of the consumers, it is necessary, or at least desirable, that somebody should keep inventories. This function can be carried out by decision-makers independent of consumers and producers. It is then natural to consider these decision-makers as "firms" in the sense of classical theory, and assume that they seek to maximize expected (utility of) profits. This will give us a theory which in many ways appears adequate. In this theory it is, however, possible that decentralized decision-making will lead to a sub-optimal situation, as indicated by our example.

REFERENCES

[1] Borch, K.: "Reciprocal Reinsurance Treaties," The *ASTIN Bulletin*, 1960, pp. 170–191.
[2] Borch, K.: "Equilibrium in a Reinsurance Market," *Econometrica*, 1962, pp. 424–444.
[3] Borch, K.: "Models for Hierarchial Decisions Involving Inventories," *Academy of Management Journal*, 1965, pp. 179–189.
[4] Lave, L.: "An Approach to the Prisoner's Dilemma," *Quarterly Journal of Economics*, 1962, pp. 424–436.
[5] Lave, L.: "Factors Affecting Co-operation in the Prisoner's Dilemma," *Behavioral Science*, 1965, pp. 26–38.
[6] Luce, R. D. and H. Raiffa: *Games and Decisions*, Wiley, 1957.
[7] Nash, J.: "The Bargaining Problem," *Econometrica*, 1950, pp. 155–162.
[8] Raiffa, H.: "Arbitration Schemes for Generalized Two-person Games," *Annals of Mathematics Studies*, No. 28, Princeton University Press, 1953, pp. 361–387.
[9] Stone, J.: "An Experiment in Bargaining Games," *Econometrica*, 1958, pp. 286–296.

Chapter XI

Elements of the General Game Theory

11.1. The simple examples which we have discussed in the two preceding chapters should have convinced the reader that game theory is the natural approach to the economics of uncertainty. Game theory is, however, far richer than our simple two-person examples can reveal. We shall therefore give a brief outline of the general theory, and we will see that a number of new aspects, of obvious economic significance, occur in games with three or more players.

In general an n-person game is described by the following three elements:

(i) A set N of n players.
(ii) n sets of strategies $S_1, S_2, \ldots, S_n$. The set S_i consists of the pure strategies $s_{i1}, s_{i2}, \ldots$ available to player i.
(iii) n payoff functions $M_1, M_2, \ldots, M_n$.

The function $M_i = M_i(s_{1r_1}, s_{2r_2}, \ldots, s_{nr_n})$ is the payoff to player i if

> Player 1 uses strategy $s_{1r_1} \in S_1$.
> Player 2 uses strategy $s_{2r_2} \in S_2$.
>
> $\cdot \quad \cdot \quad \cdot \quad \cdot \quad \cdot \quad \cdot \quad \cdot \quad \cdot$
>
> Player n uses strategy $s_{nr_n} \in S_n$.

11.2. From the minimax theorem it follows that player i can make certain that his expected payoff does not fall under a value $v(\{i\})$ regardless of what the $n - 1$ other players do.

Let us now assume that player i and player j get together and form a *coalition* which acts as one player in the game against the $n - 2$ other players. By applying a minimax strategy, this coalition can make sure that its expected payoff does not fall below a certain value $v(\{i, j\})$. It is natural to assume that the two players cannot lose by forming a coalition, so that we have

$$v(\{i, j\}) \geq v(\{i\}) + v(\{j\}).$$

To generalize this idea, let us consider an arbitrary set of players $S \subset N$. Let us assume that these players by forming a coalition can make sure that their joint payoff is at least equal to $v(S)$.

Let us next consider two coalitions, represented by two disjoint subsets R and S of N, and let $v(R \cup S)$ be the minimum gain which these two coalitions can obtain by joining forces and acting as one player.

The argument used above leads us to assume

$$v(R \cup S) \geq v(R) + v(S).$$

The function $v(S)$ is called the *characteristic function* of the game. It is a real-valued super-additive function, defined for all subsets of N. It can be proved that the whole strategic structure of the game is contained in the characteristic function.

By introducing this central concept of game theory in such a summary manner, we have swept a number of difficulties under the carpet. It has no meaning to talk about $v(S)$ as the payoff to a coalition, unless we make far-reaching assumptions about "inter-personal comparability of utility" and unlimited possibility of side-payments within coalitions. We shall, however, not formulate these assumptions. Most of the results which we shall present, can be stated without such assumptions (if we are prepared to introduce vector-valued characteristic functions and the more cumbersome notation which this will involve). The example in 10.8 gives some indications as to the shape of a more general theory.

11.3. Let us now consider an arbitrary payoff vector $x = (x_1, x_2, \ldots, x_n)$, i.e., an arrangement which gives player i the payoff x_i $(i = 1, 2, \ldots, n)$.

A payoff vector x is called an *imputation* if it satisfies the two conditions

(i) $$x_i \geq v(\{i\}) \quad \text{for all } i$$

(ii) $$\sum_{i=1}^{n} x_i = v(N).$$

These two conditions express the individual and collective rationality which we have encountered several times in the preceding chapters. It seems natural to require a payoff vector to be an imputation if it is to be seriously considered as a potential solution to a bargaining situation, at least if we look at the solution as an "arbitration scheme." However, all imputations do not have equal merits as solutions, and our problem is to eliminate some of the less attractive imputations.

Von Neumann and Morgenstern attack this problem by introducing the concept of *domination*. An imputation $y = (y_1, \ldots, y_n)$ is said to *dominate* an imputation $x = (x_1, \ldots, x_n)$ if there exists a coalition S (a subset of N) such that

(i) $$v(S) \geq \sum_{i \in S} y_i$$

(ii) $$y_i > x_i \quad \text{for all } i \in S.$$

An imputation which is dominated via some set of players S does not seem very acceptable as a solution, since these players can do better by forming a coalition and "going it alone."

Von Neumann and Morgenstern define the *solution* to the game as the set A of imputations which satisfy the following two conditions:

(i) No imputation in A is dominated by another imputation in A.

(ii) Every imputation not in A is dominated by some imputation in A.

To illustrate the meaning of this solution concept, we shall apply it to a few simple examples.

11.4. As a first illustration, let us consider a two-person game of the type we discussed in 10.8. Here the imputations are the payoff vectors (x_1, x_2) which satisfy the conditions

$$x_1 + x_2 = v(\{1, 2\}) = 5$$
$$x_1 \geq v(\{1\}) = \tfrac{3}{4}$$
$$x_2 \geq v(\{2\}) = \tfrac{6}{5}$$

Here there can be no domination, so the solution consists of *all* imputations or, in our previous terminology, of all Pareto optimal arrangements.

As our next example, let us consider a three-person game, defined by the characteristic function

$$v(\{1\}) = v(\{2\}) = v(\{3\}) = 0$$
$$v(\{1, 2\}) = v(\{1, 3\}) = v(\{2, 3\}) = 1$$
$$v(\{1, 2, 3\}) = 1$$

We shall first consider the set F consisting of the three imputations

$$(\tfrac{1}{2}, \tfrac{1}{2}, 0)$$
$$(\tfrac{1}{2}, 0, \tfrac{1}{2})$$
$$(0, \tfrac{1}{2}, \tfrac{1}{2})$$

and prove that this set is a solution.

Let us consider an arbitrary imputation (x_1, x_2, x_3). From the definition it follows that

$$\sum_{i=1}^{3} x_i = 1, \qquad x_i \geq 0.$$

It is obvious that either

(i) Two elements in this vector are equal to $\tfrac{1}{2}$, and the third is zero

or

(ii) Two elements are smaller than $\tfrac{1}{2}$,

In the former case our imputation belongs to our initial set F. In the latter case our imputation is dominated by one of the three imputations in the initial set F.

It is easy to see that none of the three imputations dominates the two others. Hence the set F constitutes a solution.

11.5. Let us next consider the set $F_1(c)$ of all imputations of the form

$$(c, x_2, x_3)$$

where c is a non-negative constant. It is obvious that

$$x_2 + x_3 = 1 - c$$

and that no imputation in $F_1(c)$ dominates any of the other imputations in the set.

Let us then consider an imputation (y_1, y_2, y_3) not in $F_1(c)$, and consider the two cases

(i) If $c < y_1$, we have $x_2 + x_3 > y_2 + y_3$. In this case we find an imputation $F_1(c)$ which dominates (y_1, y_2, y_3) by taking $x_2 > y_2$ and $x_3 > y_3$.

(ii) If $c > y_1$, we have $x_2 + x_3 < y_2 + y_3$. y_2 and y_3 cannot both be greater than $\frac{1}{2}$. If $c < \frac{1}{2}$, we an take either x_2 or x_3 greater than $\frac{1}{2}$, and hence find an imputation (c, x_2, x_3) in $F_1(c)$ which dominates (y_1, y_2, y_3).

From these considerations it follows that a solution consists of any one of the following four sets of imputations

(i) $\quad\quad\quad\quad$ F: $(\frac{1}{2}, \frac{1}{2}, 0)$, $(\frac{1}{2}, 0, \frac{1}{2})$, and $(0, \frac{1}{2}, \frac{1}{2})$

(ii) $\quad\quad\quad\quad$ $F_1(c)$: (c, x_2, x_3)

(iii) $\quad\quad\quad\quad$ $F_2(c)$: (x_1, c, x_3) $\quad 0 \le c < \frac{1}{2}$.

(iv) $\quad\quad\quad\quad$ $F_3(c)$: (x_1, x_2, c)

This means that any imputation in our simple three-person game belongs to at least one set which according to the definition of Von Neumann and Morgenstern is a *solution* to the game. This is not a very satisfactory conclusion for our purpose. We set out to find one payoff vector which we could accept as the unique solution to a bargaining situation. It is thus decidedly disturbing to find that practically every payoff vector may have a claim to be accepted as a solution.

It would, however, be premature to launch any general criticism against the solution concept of Von Neumann and Morgenstern from this basis. Most of the games which have been investigated so far have, like our simple example, an embarrassingly large number of solutions, but it has so far not been possible to prove that *every* game has a solution, i.e., that there exists in every game a non-empty set A of imputations which satisfies the two conditions in 11.3.

Von Neumann and Morgenstern argue with strength and at length that only the set A in its entirety can be considered as the solution. We shall not try to summarize these arguments, but shall endorse the following quotation from another book on game theory: "The full flavor of their argument is hard to recapture, and it can only be recommended that the reader turn to the discussion of solutions in their book" ([5] p. 206).

11.6. The essential idea behind the solution concept of Von Neumann and Morgenstern is that an imputation must have some stability properties to be acceptable as a solution. To illustrate this, let us again consider the set of imputations F which we studied in 11.4.

Let us assume that the players by some process have arrived at the payoff vector

$$(\tfrac{1}{2}, \tfrac{1}{2}, 0).$$

This is not very satisfactory to player 3, and he may approach player 2 and propose the imputation

$$(0, \tfrac{3}{4}, \tfrac{1}{4}).$$

This dominates the original imputation via coalition (2, 3).

If player 2 accepts this, player 1 will be the dissatisfied one. He may then approach player 3, and propose the imputation

$$(\tfrac{1}{2}, 0, \tfrac{1}{2})$$

which dominates $(0, \tfrac{3}{4}, \tfrac{1}{4})$ via coalition (1, 3). Player 3 will in all probability accept this.

Now player 2 may approach any of the other players and propose imputations, such as

$$(\tfrac{3}{4}, \tfrac{1}{4}, 0) \quad \text{or} \quad (0, \tfrac{1}{4}, \tfrac{3}{4}).$$

However, these two players should by now have learned by observing the fate of player 2. He was greedy and tried to obtain more than the solution allocated to him; as a result, he lost everything. Hence the imputation $(\tfrac{1}{2}, 0, \tfrac{1}{2})$ may remain stable.

11.7. In 11.3 we defined an imputation as payoff vector $x = (x_1, \ldots, x_n)$, which satisfied the two conditions

(i) $$x_i \geq v(\{i\}) \quad \text{for all } i$$

(ii) $$\sum_{i=1}^{n} x_i = v(N).$$

[154]

The first of these conditions expresses the individual rationality, i.e., that no player will accept less than he can obtain by acting alone against the $n - 1$ other players.

It is tempting to assume that each coalition will exercise the same degree of rationality as an individual player. This leads us to lay down a third condition:

(iii) $$\sum_{i \in S} x_i \geq v(S) \quad \text{for all } S \text{ contained in } N.$$

The set of all imputations which satisfy these three conditions is called the *core* of the game, a concept first introduced by Gillies [2].

At first sight the core appears to be a very attractive concept—just the answer to our search for a device which will cut down the number of imputations which may be considered as potential solutions to our problem. The trouble is, however, that the core often does this job too drastically. In a number of games the core is empty; i.e., there exists no imputation which satisfies the three conditions.

In the three-person game which we discussed in 11.4 the core is the set of vectors (x, x_2, x_3) with non-negative elements satisfying the four conditions

$$x_1 + x_2 \geq v(\{1, 2\}) \quad = 1$$
$$x_1 + x_3 \geq v(\{1, 3\}) \quad = 1$$
$$x_2 + x_3 \geq v(\{2, 3\}) \quad = 1$$
$$x_1 + x_2 + x_3 = v(\{1, 2, 3\}) = 1.$$

Adding the first three conditions, we obtain

$$x_1 + x_2 + x_3 \geq \tfrac{3}{2},$$

which obviously contradicts the fourth condition.

If the core is non-empty, i.e., if we have

$$v(\{1, 2\}) + v(\{1, 3\}) + v(\{2, 3\}) \leq 2v(\{1, 2, 3\}),$$

we can obtain the following interval for the payoff to player 1:

$$v(\{1\}) \leq x_1 \leq v(\{1, 2, 3\}) - v(\{2, 3\}).$$

The right-hand inequality states that player 1 cannot obtain more than what he contributes by joining a coalition of the two other players.

11.8. The contribution which a player can make by joining a coalition appears on intuitive reasons to be an essential element of his bargaining power. To develop this idea, let us consider an arbitrary coalition S of s players, and assume that player i, who was not in S, joins the coalition.

The contribution which player i makes to the total payoff is

$$v(S \cup \{i\}) - v(S).$$

Let us now assume that the characteristic function is such that an all-player coalition must be formed in order to obtain the maximum total payoff. Let us further assume that this n-player coalition is formed by the players joining the coalition one at the time, i.e., by a successive buildup of 2-, 3-, 4-, ... player coalitions. The n players can be ordered in $n!$ different sequences. Hence the n-person coalition can be formed in $n!$ different ways.

The s players who are in the coalition S before player i joins can be arranged in $s!$ different ways. The $n - s - 1$ players who join the coalition after player i can be arranged in $(n - s - 1)!$ different ways.

If all the $n!$ ways in which the n-player coalition can be formed are equally probable, there will be a probability

$$\frac{s!\,(n - s - 1)!}{n!}$$

that player i will join the particular coalition S.

Let us now assume that if this should happen, the payoff to player i will be exactly his contribution $v(S \cup \{i\}) - v(S)$. This means that the expected payoff to player i is

$$\varphi_i = \sum_{S \subset N} \frac{s!\,(n - s - 1)!}{n!} \{v(S \cup \{i\}) - v(S)\}$$

where the sum is taken over all subsets of N. φ_i is called the *Shapley value* of the game to player i.

The expression above was derived by Shapley [12] using an approach quite different from ours. Shapley studied the *value* which a player should assign to his right to participate in a game. He suggested that a reasonable value concept should satisfy the following three conditions:

(i) The value should be determined by the characteristic function and be independent of how we label the players.
(ii) The set of values to the n players should be an imputation.
(iii) If two games are merged into one, the value of the new game should be the sum of the two original games.

From these conditions it follows that the value must be a weighted sum of the values of the characteristic function. It may be a little surprising that these innocent looking conditions are sufficient to determine a unique set of weights. It is less surprising that the weights should be just the ones we have found. Other weights would imply that some ways of forming the n-person coalition were more probable than others.

11.9. As an illustration we shall calculate the Shapley value for a few simple games. For $n = 2$ there are only four subsets $\emptyset$, $\{1\}$, $\{2\}$, and $\{1, 2\}$, so that the Shapley value to player 1 is

$$\varphi_1 = \tfrac{1}{2}\{v(\{1, 2\}) - v(\{2\})\} + \tfrac{1}{2}\{v(\{1\}) - v(\emptyset)\}$$
$$= \tfrac{1}{2}\{v(\{1, 2\}) - v(\{2\}) + v(\{1\})\}.$$

$v(\emptyset) = 0$; i.e., the empty coalition gets a zero payoff.

For player 2 we find

$$\varphi_2 = \tfrac{1}{2}\{v(\{1, 2\}) - v(\{1\}) + v(\{2\})\}.$$

If the game is zero-sum, we have $v(\{1, 2\}) = 0$ and $v(\{1\}) = -v(\{2\})$, so that the expression reduces to

$$\varphi_1 = v(\{1\}) \quad \text{and} \quad \varphi_2 = v(\{1\}).$$

This is the minimax solution to the game.

The Nash solution to the game is the payoff vector (x_1, x_2) which maximizes the product

$$\{x_1 - v(\{1\})\}\{x_2 - v(\{2\})\}$$

subject to the condition

$$x_1 + x_2 = v(\{1, 2\}).$$

If we solve this maximizing problem, we find

$$x_1 = \varphi_1 \quad \text{and} \quad x_2 = \varphi_2.$$

For the three-person game we discussed earlier, we find

$$\varphi_1 = \tfrac{2}{6}\{v(\{1, 2, 3\}) - v(\{2, 3\})\} + \tfrac{1}{6}\{v(\{1, 2\}) - v(\{2\})\}$$
$$+ \tfrac{1}{6}\{v(\{1, 3\}) - v(\{3\})\}$$
$$= \tfrac{1}{3}.$$

Since the game is symmetric, the Shapley value of the two other players is also equal to $\tfrac{1}{3}$.

These examples show that the Shapley value contains the minimax solution and the Nash solution as special cases. This in itself should indicate that the Shapley value may have a considerable merit as the solution to our problem.

It is also worth noting that the Shapley value is equal to the average payoff of the three imputations in the set F, considered in 11.4; i.e., if these three imputations can be considered as equally probable, the expected payoff of a player is the Shapley value.

11.10. The Shapley value appears very reasonable as an *arbitration scheme*. It seems natural that an arbiter should consider the contributions which a player can make to all possible coalitions and then assign him a payoff which is a weighted average of these potential contributions. If the arbiter's decision must be accepted in advance, he does not have to worry about stability. In our example he can decide on the imputation $(\frac{1}{3}, \frac{1}{3}, \frac{1}{3})$, even if any two players can form a coalition and bring about an imputation of the type $(\frac{1}{2}, \frac{1}{2}, 0)$.

As a *predictor* the Shapley value is less attractive, mainly because it is crude. It gives us only the expected payoff, without stating anything about which payoffs actually are possible. In a sense the Shapley value is the extreme opposite of the Von Neumann–Morgenstern solution, which gives all possible payoffs without stating anything about their probability.

It would indeed be desirable to construct a theory which assigns probabilities to all imputations in the Von Neumann–Morgenstern solution. This would give us a predictive theory, which could be tested experimentally, and also an arbitration scheme, which could be used when stability conditions can be ignored. In order to develop a theory of this kind, it is clearly necessary to introduce *additional assumptions* about the behavior of the players—a point stressed by Von Neumann and Morgenstern.

To illustrate the last point, let us assume that three persons will receive $30 if they can agree—by majority vote—how the money is to be divided among them. We will not be surprised if our three players come out of the huddle announcing that they have agreed on one of the three divisions

$$(15, 15, 0), \quad (15, 0, 15), \quad \text{or} \quad (0, 15, 15).$$

If we know nothing about the personality of the players, it is natural that we should *ex ante* assign the same probability to these three outcomes.

If the players should agree on the division (10, 10, 10), we may accept this also without surprise. We will, however, be unable to say anything about the probability of this outcome compared to the probability of the three first outcomes. Should the players agree on, say (5, 10, 15), we would be surprised. If we had been asked to bet on the outcome in advance, we would probably feel cheated because we would feel that the game had been fixed or that information about the players had been withheld from us.

11.11. Our findings so far indicate that we must somehow make a choice. It seems impossible to find a solution concept which has all the desirable stability properties and also seems a fair and reasonable outcome of the game—such as an arbiter might propose.

An attempt to capture the best of both worlds has been made by Aumann and Maschler [1], who introduced the ingenious, and mathematically very elegant, concept of the *bargaining set*, which consists of

(i) The core, if the game has a non-empty core.
(ii) Some payoff vectors—not necessarily imputations—which satisfy certain stability conditions.

A number of different bargaining sets have been studied, each corresponding to a different formulation of the stability conditions. The common starting point of all these studies is a *partition* $N_1, N_2, \ldots, N_m$ of the set N of players. The partition represents a coalition structure, which together with a payoff vector $(x_1, x_2, \ldots, x_n)$ defines a *payoff configuration*. This means essentially that if, in the final phase of the game, the players are divided in the coalitions $N_1, \ldots, N_m$, playing against each other, the resulting payoff to player j is x_j ($j = 1, 2, \ldots, n$). The payoff configuration is rational if

$$\sum_{j \in N_i} x_j = v(N_i) \quad i = 1, 2, \ldots, m$$

$$x_j \geq v(\{j\}) \quad j = 1, 2, \ldots, n.$$

Let us now assume that player j in coalition R is dissatisfied with his payoff, and that he thinks player k in the same coalition gets too much. He may then consider another coalition S in the partition. If there exists a payoff vector $(y_1, \ldots, y_n)$ such that

$$y_j > x_j$$

$$y_s \geq x_s \quad \text{for } s \in S$$

$$y_j + \sum_{s \in S} y_s = v(S \cup \{j\}),$$

it seems that player j can increase his payoff by leaving the coalition R and joining S. We say that he has an *objection* against player k.

Player k may, however, be able to hold his own by leaving R and joining another coalition T in the partition. This will be the case if there exists a payoff vector $(z_1, \ldots, z_n)$ such that

$$z_k \geq x_k$$

$$z_t \geq x_t \quad \text{for } t \in T$$

$$z_t \geq y_t \quad \text{for } t \in S \cap T$$

$$z_k + \sum_{t \in T} z_t = v(T \cup \{k\}).$$

In this case we say that he has a *counterobjection* to the objection of player j. A rational payoff configuration is called *stable* if we can find a counterobjection to each objection. The bargaining set is the set of all stable rational payoff configurations.

We have given a very heuristic presentation of the main ideas behind the bargaining set, so it may be useful as an illustration to analyze the three-person game in 11.7. Let us consider a payoff configuration consisting of

> The coalition structure $\{1, 2\}, \{3\}$
> The payoff vector (x_1, x_2, x_3)

The rationality conditions are

$$x_1 + x_2 = v(\{1, 2\})$$

$$x_3 = 0.$$

Player 1 will have an objection against player 2 if there exists a vector (y_1, y_2, y_3) such that

$$y_1 > x_1$$

$$y_3 \geq 0$$

$$y_1 + y_3 = v(\{1, 3\}).$$

Player 2 will have a counterobjection if there exists a vector (z_1, z_2, z_3) such that

$$z_2 \geq x_2$$

$$z_3 \geq y_3$$

$$z_2 + z_3 = v(\{2, 3\}).$$

We have here a system of linear equations and inequalities. We can construct a similar system corresponding to the counterobjections which player 1 has against the objections of player 2. The bargaining set consists of the payoff configurations $(x_1, x_2, x_3, \{1, 2\}, \{3\})$ for which these systems have non-negative solutions in the y's and z's.

We can verify by elementary arithmetics that

$$x_1 = v(\{1, 2\}) \quad \text{for } v(\{1, 3\}) > v(\{1, 2\}) + v(\{2, 3\})$$

$$x_1 = 0 \quad \text{for } v(\{2, 3\}) > v(\{1, 2\}) + v(\{1, 3\})$$

$$v(\{1, 3\}) \leq x_1 \leq v(\{1, 2\}) - v(\{2, 3\})$$
$$\text{for } v(\{1, 2\}) > v(\{1, 3\}) + v(\{2, 3\})$$

$$x_1 = \frac{v(\{1, 2\}) + v(\{1, 3\}) - v(\{2, 3\})}{2} \quad \text{otherwise.}$$

The payoff vector is then determined by the conditions $x_2 = v(\{1, 2\}) - x_1$ and $x_3 = 0$.

We can carry through the same calculations for the other partitions, and we can show that the core will be contained in the bargaining set. For the special game considered in 11.4, we will find that the bargaining set consists of the set F and the payoff vector $(0, 0, 0)$ corresponding to the coalition structure $\{1\}, \{2\}, \{3\}$.

The last allocation has obviously no place in a normative theory or an "arbitration scheme." In a theory which tries to predict the outcome of a bargaining situation, it is only appropriate to consider the possibility that the players may be unable to reach any kind of agreement.

Aumann and Maschler do not offer any suggestions as to the relative importance of the different elements of the bargaining set. This would, however, hardly be possible without making additional assumptions about the environment in which the bargaining takes place.

A series of experiments by Maschler [6] with high school students in Jerusalem indicates that the bargaining set gives a fairly good prediction of the outcome of some bargaining situations.

We shall not discuss these experiments any further, since it is more convenient to discuss another experiment by Selten and Schuster [11], which can be considered as a follow-up of the work by Maschler.

11.12. A number of authors have suggested "additional assumptions" about the behavior of the players in order to obtain a unique solution to the game. It is not possible to discuss all these suggestions in the present context, even if we restrict ourself to the behavioral assumptions which are clearly relevant to economic situations. It may, however, be useful to refer to some of the papers by Harsanyi [3] and [4], and to try to outline the main ideas behind his work.

The argument which led to the Nash solution [7] for the two-person game can obviously be generalized to n-person games. It will then give as solution the payoff vector $(x_1, \ldots, x_n)$ which maximizes the product

$$[x_1 - v(\{1\})][x_2 - v(\{2\})] \cdots [x_n - v(\{n\})]$$

subject to the conditions

$$x_1 + x_2 + \cdots + x_n = v(N)$$

$$x_i \geq v(\{1\}) \quad (i = 1, 2, \ldots, n).$$

It is easily verified that this vector is given by

$$x_i = v(\{i\}) + \frac{1}{n} v(N) - \frac{1}{n} \sum_{i=1}^{n} v(\{i\}).$$

This solution depends only on the values of the characteristic function for the coalitions consisting of one player and of all n players; i.e., it ignores the advantages which players can obtain by forming intermediary coalitions during the bargaining. This is unrealistic, but it is a good starting point.

Initially it is natural to assume that player i considers only $v(\{i\})$, i.e., the payoff he can secure for himself if he should be left out and have to play alone against the others. If, however, he succeeds in forming an intermediary coalition with other players, he may secure a higher payoff for himself, even if this coalition in the end should have to play alone against the others. The players in this intermediary coalition will then bargain further, but with a higher payoff requirement than when they acted alone. By formalizing this bargaining behavior, Harsanyi obtains a unique solution to the game—closely related to the Shapley value.

11.13. Most of the "additional assumptions" which have been proposed appear reasonable, and all of them have strong advocates. It therefore does not seem very useful, at the present stage, to decide on the relative merits of the various assumptions by theoretical arguments. It may be more fruitful to study the experimental evidence as to how players actually have behaved in different games played under laboratory conditions. There is an extensive literature on so-called experimental games. The most representative survey article appears still to be a paper by Rapoport and Orwant [10], which covers work done up to the end of 1961.

To illustrate the usefulness of the experimental approach, we shall give a brief account of an experiment made by Selten and Schuster [11] with students at the University of Frankfurt.

11.14. For their experiment Selten and Schuster used a five-person "weighted majority game." In such games there are only two kinds of coalitions:

(i) If $v(S) = v(N)$, S is a *winning* coalition.
(ii) If $v(S) = 0$, S is a *losing* coalition.

In the game used for the experiment, the winning coalitions were

(i) Player 1 and at least one of the other players.
(ii) Players 2, 3, 4, and 5.

Player 1 is obviously in a stronger position than the other players. If he can secure the cooperation of just one of the others, the two will get the whole payoff and can divide it between themselves. Player 1 gets nothing only when all the four other players form a coalition against him.

TABLE 11
Final allocations in 12 five-person games

Game	Allocation				
1	25,	0,	0,	15,	0
2	25,	0,	15,	0,	0
3	25,	0,	0,	0,	15
4	25,	0,	15,	0,	0
5	0,	8.66,	8.66,	8.66,	14
6	0,	7,	7,	19,	7
7	18,	0,	0,	0,	22
8	25,	0,	5,	5,	5
9	28,	0,	0,	12,	0
10	20,	0,	20,	0,	0
11	25,	0,	15,	0,	0
12	8,	8,	8,	8,	8

As total payoff Selten and Schuster took $v(N) = 40$ DM—about \$10. The game was played 12 times, each time with different players. As soon as the players had formed a winning coalition, with an agreed payoff, this was registered. If this coalition remained stable for 10 minutes of continued bargaining, it was taken as final. The allocations which the players finally agreed upon are given in Table 11.

11.15. The results in Table 11 are in many ways surprising, and they may be taken as a blow to the ability of game theory to predict the outcome of a real-life bargaining situation. The core of this game is empty, but the game has a large number of solutions. The so-called main solution consists of the five allocations

$$(30, 10, \ 0, \ 0, \ 0)$$

$$(30, \ 0, 10, \ 0, \ 0)$$

$$(30, \ 0, \ 0, 10, \ 0)$$

$$(30, \ 0, \ 0, \ 0, 10)$$

$$(\ 0, 10, 10, 10, 10).$$

This corresponds to the solution F which we discussed in 11.4. It is a solution which places the whole emphasis on stability, without any regard to the "fairness" of the allocation. It is remarkable that none of the allocations in the main solution turned up in the experiment.

The Shapley values (see 11.8) of this game are

$$\varphi_1 = 40\left\{\frac{4!}{5!} + \frac{4 \times 3!}{5!} + \frac{4 \times 3 \times 2!}{5!} + \frac{4!}{5!}\right\} = 24$$

$$\varphi_i = 40\left\{\frac{3!}{5!} + \frac{3!}{5!}\right\} = 4 \quad i = 2, 3, 4, 5$$

i.e., the allocation $(24, 4, 4, 4, 4)$. Nothing similar to this occurred in the experiment.

The bargaining set (see 11.11) allocates a payoff between 20 and 30 to player 1, and between 10 and 20 to his partner. As allocations of the form $(25, 15, 0, 0, 0)$ occurred in 5 out of the 12 games, the bargaining set seems to have some predictive power. The set, however, also contains the allocation $(0, 10, 10, 10, 10)$, and this did not turn up in the experiment.

11.16. Selten and Schuster kept detailed records of the actual bargaining processes in their experiment. This opens the way to a number of comments and intriguing speculations, of which we shall just give a small sample.

In *Game 12* it took the players 23 minutes to agree upon the allocation $(0, 10, 10, 10, 10)$. This "intermediary" coalition lasted for 4 minutes and 30 seconds before it switched to $(8, 8, 8, 8, 8)$, which remained stable for 10 minutes.

It is not easy to explain this result in terms of game theory. It seems that either the players as a group had a strong feeling of solidarity or player 1 was able to convince the others that the only fair principle was "To everyone according to his needs—and we all have equal needs." An alternative explanation may be that these players really needed more time to grasp the true nature of the game.

Game 8 began with the rather curious allocation (0, 17, 7.66, 7.66, 7.66), which was reached after 6 minutes and 20 seconds. Then followed (20, 20, 0, 0, 0) after 4 minutes and 55 seconds and (25, 0, 5, 5, 5) after 6 minutes and 55 seconds; this last allocation remained stable for 10 minutes.

It is hard to explain this outcome unless we assume that players 3, 4, and 5 somehow wanted to punish player 2 for his behavior earlier in the game. It is, however, not easy to formulate such assumptions and bring them into a mathematical model.

Game 5 ran as follows:

       (0, 10, 10, 10, 10)   after 10 minutes and 30 seconds;
       (25, 0, 0, 0, 15)   after 50 seconds;
       (0, 8.66, 8.66, 8.66, 14)   after 3 minutes and 55 seconds.

The only explanation of this outcome seems to be that players 2, 3, and 4 felt that player 5 was entitled to a great payoff because he had the good luck of getting into a coalition with the strong player 1 early in the game.

11.17. The experiment we have discussed indicates how exceedingly complex the real world may be. It is clear that, in spite of all its refinements, game theory today is far too simple to give a satisfactory analysis of the bargaining situations which we find in economic life.

The Solution—with capital *S*—to a game should give us a probability distribution over the set of possible outcomes. However, a solution of this kind cannot be derived from a mere description of the game and its rules. We must have far more information, information about the relations among the players—the network of feelings of friendship and hostility which exists in any community.

Just to describe this network and measure the strength of the different affinities is a formidable task, but one which must be accomplished if we are ever to obtain The Solution.

REFERENCES

[1] Aumann, R. and M. Maschler: "The Bargaining Set for Cooperative Games," *Annals of Mathematics Studies*, No. 52, Princeton University Press, 1964, pp. 443–476.
[2] Gillies, D.: "Solutions to General Non-zero-sum Games," *Annals of Mathematics Studies*, No. 40, Princeton University Press, 1959, pp. 47–85.

[3] Harsanyi, J.: "A Bargaining Model for the Cooperative *n*-person Game," *Annals of Mathematics Studies*, No. 40, Princeton University Press, 1959, pp. 325–355.

[4] Harsanyi, J.: "A Simplified Bargaining Model for the *n*-person Cooperative Game," *International Economic Review*, 1963, pp. 194–220.

[5] Luce, R. D. and H. Raiffa: *Games and Decisions*, Wiley, 1957.

[6] Maschler, M.: "Playing an *n*-person Game—an Experiment," Research Memorandum No. 73, Econometric Research Program, Princeton University, 1965.

[7] Nash, J.: "The Bargaining Problem," *Econometrica*, 1950, pp. 155–162.

[8] Neumann, J. von and O. Morgenstern: *Theory of Games and Economic Behavior*, Princeton University Press, 1944.

[9] Raiffa, H.: "Arbitration Scheme for Generalized Two-person Games," *Annals of Mathematics Studies*, No. 28, Princeton University Press, 1953, pp. 361–387.

[10] Rapoport, A. and C. Orwant: "Experimental Games: A Review," *Behavioral Science*, 1962, pp. 1–37.

[11] Selten, R. and K. Schuster: "Psychological Variables and Coalition Forming Behavior," in *Risk and Uncertainty*, MacMillan & Co., 1968.

[12] Shapley, L. S.: "A Value for *n*-person Games," *Annals of Mathematics Studies*, No. 28, Princeton University Press, 1953, pp. 307–317.

Chapter XII

The Objectives of the Firm

12.1. In this chapter we shall study the behavior of a particular class of decision-makers which it is convenient—and conventional—to call "the firm." In 8.10 we indicated how a general model could be constructed by assuming that the economy was made up of two classes of decision-makers:

(i) *Consumers* who seek to maximize their utility.

(ii) *Producers*, or "firms," seeking to maximize their profits.

The problem of the firm can then be formulated as follows: Given

(i) A production function

$$y = f(x_1, \ldots, x_n).$$

This function determines the amount of output y obtained by using the amounts $x_1, \ldots, x_n$ of the possible inputs (factors of production).

(ii) A set of prices q and $p_1, \ldots, p_n$ of output and inputs respectively.

If the firm decides to use an input vector $(x_1, x_2, \ldots, x_n)$, it will make a profit

$$P = qf(x_1, \ldots, x_n) - \sum_{i=1}^{n} p_i x_i.$$

The problem is then to determine the input vector which maximizes P.

The solution to this problem is the vector $(x_1, \ldots, x_n)$ determined by the equations

$$q \frac{\partial f}{\partial x_i} = p_i \quad (i = 1, 2, \ldots, n).$$

This result is given in most textbooks of economics, and it is often expressed as a theorem: *Profit is maximized when marginal revenue is equal to marginal cost.*

12.2. The model we have outlined does of course represent an enormous simplification of the real world, i.e., of the complex of problems which a real life business firm has to solve. The most obviously unrealistic aspects of our model are

(i) The production function itself is an abstraction. It may not exist except in the form of some stochastic relationship between inputs and outputs. (Any farmer, and probably most engineers, will know from bitter experience that it is not possible to predict the exact output of a production process.)

[166]

(ii) Even if a well-defined production function should exist, it may not be completely known to the management, which makes the decisions on behalf of the firm.

(iii) It is not certain that prices exist, and are independent of the quantities which the firm decides to buy or sell. Even if such prices exist, they may not be known to the firm, except in a stochastic sense.

These points are so obvious that they have been noticed by practically every economist who has made a serious study of the subject. However, few of these economists seem to have made any real attempt to generalize the naive model and give a more realistic description of the situation in which the firm makes its decisions. Instead of establishing a theoretical framework based on more realistic assumptions about the firm's *environment*, most economists seem to have chosen to criticize the assumption that the firm seeks to maximize profits. There has in fact been a veritable competition among economists in suggesting motives, apart from profit maximization, which determine the decisions made by firms.

12.3. This is in a way a surprising development. To common sense it appears almost self-evident that the objective of a businessman is to make the highest profit—one might even be tempted to take this as a definition, and say that a person who pursues different objectives is no real businessman. It seems obvious that, if a businessman has the choice of two actions leading to outcomes which are identical in every respect except that one action will give a higher profit than the other, he will choose the most profitable one. In classical theory this was in fact taken as obvious, and it seems to have been accepted, almost without question, that firms were seeking to maximize profits.

We must of course admit that the decisions which a businessman has to make are rarely, if ever, so simple and clear cut that they consist in selecting the most profitable among a few well-defined actions. This should, however, indicate that we ought to study the *environment* in which the decisions are made rather than the objectives behind the decisions. The environment will usually be made up of other decision-makers, and their decisions will create the situation in which our particular firm has to make its decision. This inevitably leads to a game-theoretical situation, and if mixed strategies enter, there will be uncertainty about the outcome of the game.

12.4. Economists who contest the importance of the profit motive in business decisions do, of course, have their reasons. Those most frequently quoted appear to be the following:

(i) When decisions actually made by a firm are analyzed, it may turn out that there were other possible decisions which would have given a higher profit. From such observations we can simply conclude that the firm has made the "wrong" decisions. We can also take a more sophisticated attitude and construct a theory in which decisions made *ex ante* do not come out as expected when analyzed *ex post*. However, the easiest, and apparently most fashionable, way out of the dilemma is to assume that the firm did not in the first place try to maximize profits. This way may be the most agreeable explanation to a top manager, since it contains no reference to mistakes or to his possible fallibility.

(ii) When businessmen are interviewed by research economists, they will usually state that there are a number of elements other than profits which are considered when business decisions are made. They may, however, not be very specific as to what these elements are. Businessmen may make such statements because they feel that there is something unethical about having profits as one's sole motive, but this is probably not a very important element. In all fairness we should assume that businessmen are sincere, although not always very articulate, when they try to explain their objectives to a research economist.

12.5. To get a firmer grip on the elements which enter into consideration when business decisions are made, let us consider a farmer who has grown a number of different crops on his land during the last year. Let us assume that at Christmas time this farmer shows the farm accounts to his son, who is home from a business school during the vacation. Having examined the account as an expert and heir, the boy may well exclaim, "Dad, you would have made much more money if you had planted potatoes on all our land!"

The farmer may make any number of unprintable answers, but he may also reply along one of the following lines:

(i) "I like to see cornfields round the house." This implies that some *non-monetary* elements are considered when the farmer makes his decisions.

(ii) "I did not know that it would be such a bumper year for potatoes." This indicates that there are some *uncertainty* elements which the farmer felt he had to take into consideration. He did not want to put all his eggs in one basket.

(iii) "Our land will soon be exhausted, Son, unless I rotate the crops." This means that there is a *time* element in the farmer's decisions. The son may then recall that his textbooks were not always quite clear when they discussed short-term and long-term profits, and he may become a little uncertain as to what his father should maximize.

Each of these replies suggests a class of objectives which may be found behind decisions, not only in farming, but in any economic activity. We shall discuss these classes, and we shall argue that most of the objectives of firms which have been suggested in economic literature fit into one of these three classes.

12.6. *Non-monetary* elements undoubtedly play some part in business decisions, but they may not be quite as important as some authors have suggested. If a firm states that one of its principal objectives is to maintain a "good image with the public," this may represent a genuine non-monetary objective. The manager may enjoy the good image of his firm, for instance, because it makes him a respected and popular member of his community. However, a cynic may well argue that a good manager is unlikely to sacrifice immediate profits in order to build up a good image unless this good image can be turned into higher profits at a later date. If the cynic is right, this means that the apparently non-monetary objective can be explained as a profit objective when the time element is brought into the model.

In large corporations, decisions may be made by employees (executives) whose renumeration is practically independent of the profits earned by the corporation. This may give the executives a motivation for making decisions which an outsider can only interpret as a pursuit of some non-monetary objective on behalf of the firm. This raises some problems which we shall return to in the next chapter.

The *uncertainty* element was discussed in considerable detail in earlier chapters, and we shall now indicate how this element can be brought into the theory of the firm.

The *time* element is obviously very important. This should be evident even from the brief discussion in the preceding paragraphs. We shall return to this problem and analyze it in more detail in Chapter XIII.

12.7. At this stage it may be useful to discuss some of the alternatives to profit maximizing which have been seriously suggested in the literature.

Baumol ([1] p. 192) studies the following two objectives, which have also been discussed by other authors:

(i) The firm seeks to *maximize sales*, subject to profits staying above some acceptable minimum. In the notation of 12.1 this means that the firm will seek

$$\max \{qf(x_1, \ldots, x_n)\}$$

subject to

$$qf(x_1, \ldots, x_n) - \sum_{i=1}^{n} p_i x_i \geq M$$

where M is the acceptable minimum.

This can be interpreted as a non-monetary objective if the manager wants a high volume of sales simply because it makes him feel important.

However, high sales may imply that the firm will maintain a strong competitive position in the future, and that this in due time will pay off in the form of high profits. Hence the real objective of the firm may well be to maximize long-term profits, defined in a suitable manner.

(ii) The firm seeks to *minimize unit cost*. This leads to the mathematical problem of determining

$$\min \left\{ \frac{p_1 x_1 + \cdots + p_n x_n}{f(x_1, \ldots, x_n)} \right\}.$$

This may be a non-monetary objective if it has been established by engineers who take a particular pride in technical perfection for its own sake. However, low unit cost means that the firm will be in a strong position if the price of its output should fall. Hence this objective can also be interpreted as a profit objective, if uncertainty is taken into consideration.

12.8. We shall not attempt to discuss all the different objectives of the firm which have been proposed by various authors. It may, however, be useful to say a few words about *liquidity* considerations, since they have played an important part in the literature, and since they are particularly relevant to the main subject of this book.

The following statement by Dean ([3] p. 32) is typical:

As a consequence of balance-sheet considerations, executives sometimes deliberately choose a less profitable but more liquid alternative, when as is common, the two pull in opposing directions.

To illustrate, let us consider the following alternatives:

(i) Sink all your resources into a prospect which is expected to pay $100,000 some time next year.
(ii) Keep a liquid reserve of $30,000, and put the rest of your resources into a prospect which is expected to pay $50,000 next year.

Dean's statement simply means that he has sometimes observed executives in such situations to prefer decision (ii). Most of us have probably made similar observations, without any surprise. It is, however, difficult to explain such decisions without assuming that the situation contains some elements of uncertainty. If a firm keeps on hand an amount of cash which is not needed for some specific purpose, we must assume that the firm feels that the cash may be needed for unspecified contingencies, or to make use of favorable opportunities that might turn up.

It seems that liquidity considerations ought to be seen as a part of an overall profit objective when we consider time and uncertainty. It is, however, possible that a handsome amount of cash will give a pleasant feeling of security which can hardly be justified on rational grounds. The desire for liquidity may then become almost a "non-monetary" element—in spite of the contradiction in terms.

12.9. The examples we have discussed in the two preceding sections can be interpreted as *rules of thumb*. The overall objective of a corporation may be very complex. It may be difficult, and not worthwhile, for top management to spell it out in detail to junior executives—or to interviewing research economists. We should also realize that it may be an advantage to a firm to keep its real objectives secret. If we know all about the objectives of a firm which seeks to buy something from us, we will also know how much the firm is prepared to pay. We can then insist on this amount, regardless of the lower offers which the firm may make in an attempt to strike a bargain. It appears particularly important to keep the objectives secret in reinsurance negotiations. This problem has been discussed in another paper [2], which also gives some examples.

For practical purposes it may be sufficient to give the junior executives simple and clear instructions. If they are told, for instance, to maximize sales, subject to certain restrictions on expenditure, they will have a well-defined and challenging task. If the junior executives handle these tasks well, they will help the corporation to achieve its overall objective, which may be quite unknown to them.

It seems clear that none of the objectives we have discussed can have any general validity. It is possible that a small manufacturer will prefer:

Situation 1: Sales: $110,000
Profits: $19,000

to

Situation 2: Sales: $100,000
Profits: $20,000.

We will, however, expect him to consider

Situation 3: Sales: $99,000
Profits: $50,000

as far more attractive than the two first situations.

Behind the assumption that firm seeks to maximize sales, there is probably a tacit assumption that something like situation 3 will never be available.

12.10. Let us now return to the model of 12.1 and introduce some uncertainty. For the sake of simplicity we shall assume that there is uncertainty only about the price of the output; i.e., we shall assume that q is a stochastic variable with a distribution $G(q)$. We shall also assume that a density function $g(q) = G'(q)$ exists. These assumptions imply that the profit

$$P = qf(x_1, \ldots, x_n) - \sum_{i=1}^{n} p_i x_i = qf(x) - p'x$$

is also a stochastic variable. It will then make no sense to talk about maximizing profits, and this is the real reason why profit maximization has to be rejected as the objective of a firm. It simply has no meaning when uncertainty is introduced.

To each input vector $x = (x_1, \ldots, x_n)$ which the firm may choose in this situation, there will correspond a *profit distribution*. The objective of the firm must then be to find the input vector which gives the "best" of the attainable profit distributions. In order to solve this problem, the firm must, as we have seen in Chapter III, have a preference ordering over the set of attainable profit distributions. If this preference ordering is consistent, it can be represented by a utility function $u(P)$. The problem of the firm can then be reduced to determining the input vector which maximizes

$$U = \int_0^\infty u[qf(x) - p'x]g(q)\, dq.$$

The first-order conditions for a maximum are

$$\frac{\partial U}{\partial x_i} = \int_0^\infty u'[qf(x) - p'x]\left\{ q\frac{\partial f}{\partial x_i} - p_i \right\}g(q)\, dq = 0, \quad (i = 1, 2, \ldots, n).$$

For reasonably well-behaved utility functions these equations will have as solution a vector $\hat{x} = \{\hat{x}_1, \ldots, \hat{x}_n\}$, and there will exist a price $\hat{q}$ such that

$$\hat{q}\left[\frac{\partial f}{\partial x_i}\right]_{x = \hat{x}} = p_i.$$

This means that the firm will act *as if* it was certain that the price of output would be $\hat{q}$; i.e., the price $\hat{q}$ serves as a "certainty equivalent," which determines the decision of the firm.

12.11. In general $\hat{q}$ will be different from the price which the firm actually obtains for its output. This may, as we mentioned in 12.4, lead an observer to conclude that the objective of the firm is not to maximize profits. This is of course trivially true, since profit maximizing becomes meaningless once uncertainty has been introduced in the model.

In economic theory one has often "assumed away" uncertainty by replacing stochastic variables with their expected values; i.e., in our example one would assume that the firm acted as if it was certain that the price would be

$$\bar{q} = \int_0^\infty qg(q)\, dq.$$

If the utility function is concave, we will have

$$\bar{q} > \hat{q}.$$

This may again lead observers to conclude either that the firm makes the "wrong" decisions or that the objective of the firm is not to maximize expected profits. These observers may then construct hypotheses as to the relative weights which the firm attaches to the various other possible objectives, and test these hypotheses against observed behavior in business. Such research can lead to hypotheses which fit the data reasonably well, and which may be useful in practice.

This approach may lead to a long list of more or less interrelated objectives, which together determine the decisions of the firm, and this is a complicated and hence not a very attractive theory. We get a much simpler and more satisfactory theory if we assume that, as a first approximation, the sole concern of the firm is profits, and that it seeks to obtain the best of the available profit distributions.

12.12. The example we have discussed in the preceding two sections is in many respects very artificial. It may, therefore, be useful to discuss a more realistic example.

Let us consider an insurance company with an initial capital S, and assume that the company receives an amount P by underwriting a portfolio of insurance contracts. Let us further assume that the company will have to pay an amount x to settle claims made under the contracts in the portfolio, and that x is a stochastic variable with the distribution $F(x)$. This transaction will give the company the expected utility

$$U(1) = \int_0^\infty u(S + P - x)\, dF(x)$$

where $u(x)$ is the utility function which represents the company's attitude to risk.

If the company is not forced into this transaction, we must have

$$u(S) < \int_0^\infty u(S + P - x)\, dF(x);$$

i.e., the transaction must give the company an increase in utility.

Let us now assume that the company can *reinsure* a quota $1 - k$ of the portfolio on "original terms." This means that the company pays an amount $(1 - k)P$ to a reinsurer, who in return undertakes to pay an amount $(1 - k)x$ if total claims are x. This arrangement means that the company *retains* a quota k of the portfolio which it has underwritten. The reinsurance arrangement will give the company the expected utility

$$U(k) = \int_0^\infty u(S + kP - kx)\, dF(x).$$

The decision problem of the company is then to determine the value of k which maximizes $U(k)$, subject to the condition $0 \le k \le 1$.

In this model the company cannot conceivably have any other objective than to maximize expected utility. The model is extremely simple, but it gives an adequate representation of some real-life situations. There are insurance companies, for instance those serving members of a closed group, which accept new business only at the beginning of the calendar year, and which issue only contracts of one year's duration. When a company of this kind closes its books on January 2, its only objective is to find the best possible reinsurance arrangement. Profit maximizing has no meaning to this company.

The objective of the company may be to maximize *expected* profits. This means that the company's utility function is linear. It is easy to see that in this case the company will not reinsure at all, so that the solution is $k = 1$, provided of course that

$$P > \int_0^\infty x \, dF(x);$$

i.e., that the portfolio was actuarially favorable to the company.

The fact that insurance companies usually reinsure, proves that their objective is not to maximize expected profits.

12.13. As another example let us consider a firm with the production function

$$y = (x_1 x_2)^{1/2}.$$

Let the prices be q, p_1, and p_2. If the firm uses the input vector (x_1, x_2), the profit will be

$$P_1 = q(x_1 x_2)^{1/2} - p_1 x_1 - p_2 x_2 = qy - p_1 x_1 - p_2 x_2.$$

Let us for the sake of simplicity assume that demand y is given and constant, so that the problem is to determine the input vector which will enable the firm to produce y at the lowest possible cost. It is easy to see that cost is minimized for

$$x_1 = y\left(\frac{p_2}{p_1}\right)^{1/2} \quad \text{and} \quad x_2 = y\left(\frac{p_1}{p_2}\right)^{1/2}$$

and hence that maximum profit is

$$P'_1 = y\{q - 2(p_1 p_2)^{1/2}\}.$$

Let us now interpret the first input as labor and the second as capital, and assume that the firm expects that it will receive orders to produce the output y in t successive periods. Let us further assume that the capital equipment will be without value after t periods, because there will be no more demand for the output which can be produced with this equipment.

If the input vector (x_1, x_2) is used, total profits over the t periods will be

$$P_t = tq(x_1 x_2)^{1/2} - tp_1 x_1 - p_2 x_2 = tqy - tp_1 x_1 - p_2 x_2$$

where $x_1 x_2 = y^2 = $ a given constant.

of the Firm

We find that the maximum profit is

$$P'_t = y\{tq - 2(tp_1p_2)^{\frac{1}{2}}\},$$

which is obtained for

$$x_1 = y\left(\frac{p_2}{tp_1}\right)^{\frac{1}{2}} \quad \text{and} \quad x_2 = y\left(\frac{tp_1}{p_2}\right)^{\frac{1}{2}}.$$

We see that an increase in t will lead to a decrease in x_1 and to an increase in x_2, i.e., to a more capital-intensive production process.

12.14. To introduce some uncertainty, let us now assume that the firm does not know how long the demand for the output y will last.

Let $f(t)$ be the probability that demand will last for exactly t periods. We shall assume

$$\sum_{t=1}^{\infty} f(t) = 1.$$

The input vector (x_1, x_2) will then give the firm an expected profit

$$\bar{P} = \sum_{t=1}^{\infty} P_t f(t) = qy \sum_{t=1}^{\infty} tf(t) - p_1 x_1 \sum_{t=1}^{\infty} tf(t) - p_2 x_2$$

or

$$\bar{P} = \bar{t}qy - \bar{t}p_1 x_1 - p_2 x_2$$

where

$$\bar{t} = \sum_{t=1}^{\infty} tf(t)$$

is the expected duration of the demand.

By comparing this expression with results in 12.13 we see that the firm will maximize *expected* profits if it takes it for certain that demand will last for $\bar{t}$ periods, and then maximizes profits. This policy may lead to heavy losses if $\bar{t}$ is large, since if demand should last only for a short time, the firm may be caught with a large amount of worthless capital equipment. If this possibility worries the firm, it has a "risk aversion." In order to formalize this, we can assume that the objective of the firm is to maximize expected utility with a basic utility function which is concave. This leads us to the problem of determining

$$\max_{x_1 x_2 = y^2} \left\{ \sum_{t=1}^{\infty} u(tqy - tp_1 x_1 - p_2 x_2) f(t) \right\}.$$

If the firm solves this problem and acts accordingly, an outside observer may well conclude that the firm is too pessimistic about the duration of the demand for its product. He may, however, also conclude that the firm operates in an inefficient manner because it has not been willing to take the risk of investing enough in fixed capital equipment.

[175]

12.15. The simple model which we have discussed in the two preceding sections can obviously be generalized. One way of doing this is to assume that demand in period t, y_t $(t = 1, 2, \ldots)$, is a known stochastic process. We can then assume that the firm has to choose a certain capitalization x_2, which cannot be changed during the next n periods.

If this model is worked out in detail, it will offer some very interesting mathematical problems, but since the model itself is not very satisfactory we shall not discuss them. In this more general model it does not appear natural to assume that the firm should be concerned only with the probability distribution of the total amount of profits earned before the capital equipment becomes obsolete. If there is a possibility that this equipment may serve for a long time, it may make a considerable difference whether the profit is earned early or late during the lifetime of the equipment. This indicates that we must introduce a *time element* in our models if we are to come to grips with the real problems.

The usual approach to the time problem is to assume that an early payment is preferred to a late one and "discount" a delayed payment at a certain interest rate r. The *discount factor* v is defined as

$$v = \frac{1}{1 + r} \leq 1,$$

and the *cash value* of an amount x, payable after t periods, is $v^t x$.

Let us now consider a sequence of payments $x_1, \ldots, x_t, \ldots, x_n$, where x_t is the amount payable after t periods. The cash value of this sequence is then

$$\sum_{t=1}^{n} v^t x_t.$$

If all payments are equal, i.e., $x_t = x$ for all t, the sum reduces to

$$\sum_{t=1}^{n} v^t x = v \frac{1 - v^n}{1 - v} x.$$

Going back to the formulae in 12.14, we find that the input vector (x_1, x_2) will give the firm an expected discounted profit

$$\bar{P}_v = qy \sum_{t=1}^{\infty} \frac{v}{1 - v} (1 - v^t) f(t)$$

$$- p_1 x_1 \sum_{t=1}^{\infty} \frac{v}{1 - v} (1 - v^t) f(t) - p_2 x_2$$

or

$$\bar{P}_v = \bar{\imath}_v qy - \bar{\imath}_v p_1 x_1 - p_2 x_2$$

where

$$\bar{\imath}_v = \frac{v}{1 - v} \left\{ 1 - \sum_{t=1}^{\infty} v^t f(t) \right\}.$$

A reasonable objective of the firm may be to maximize $\bar{P}_v$ at a suitable discount rate. It is easy to show that $\bar{\iota}_v \leq \bar{\iota}$, and that discounting will lead the firm to a less capital-intensive production process.

12.16. Let us now consider the more general problem of selecting the best of two sequences of payments

$$x = \{x_1, x_2, \ldots, x_t, \ldots\}$$

and

$$y = \{y_1, y_2, \ldots, y_t, \ldots\}.$$

Formally this is a decision problem under certainty of the type we discussed in Chapter II. To solve the problem, we need a preference ordering over the set of all such sequences. If this ordering can be represented by a utility function $u(x)$, our decision rule is to select x in preference to y if and only if $u(x) > u(y)$.

If we consider only finite sequences—of, say, n terms—the utility function will be an ordinary function of n variables; i.e., $u(x) = u(x_1, \ldots, x_n)$. We can then analyze the problem with the methods of classical economic theory by simply interpreting payments at different times as different commodities. This general formulation may appear useless, since the only utility functions which have found any practical application are of the form

$$u(x_1, \ldots, x_n) = \sum_{t=1}^{\infty} v^t x_t.$$

These functions imply discounting at a constant rate, and it seems desirable to study timing preferences of a more general nature. As an illustration, let us study the following two sequences of payments:

(i) $\{3, 4, 1, 3, 2, 2, 3\}$

(ii) $\{1, 1, 1, 2, 2, 2, 3\}.$

At any rate of discount, (i) will appear as the better of the two sequences. This is a natural preference ordering if the sequences are interpreted as the returns from two investments which are offered us at the same price.

If, however, we interpret the sequences as the profits a firm will make in successive periods, it is conceivable that management will prefer the decision which leads to sequence (ii). This "earning record" may create the impression that the firm is under responsible management and has achieved steady growth.

If the two sequences are interpreted as national income available for consumption under different development plans, it is also possible that (ii) will be preferred for political reasons.

[177]

These considerations indicate that preference orderings over sequences of payments should, in some contexts, contain an element which we can call "a desire for stability." If we establish preference orderings by simple discounting, this element will escape us.

It is worth noting that in certain cases it may be desirable to build a "desire for change" into a preference ordering. During a visit to Paris, Wold and some of his Swedish friends found French wines more interesting than the good cheap milk readily available in Sweden. They agreed, however, that if they had to choose either wine or milk as their only beverage for the rest of their lives, they would—possibly reluctantly—choose milk [7].

12.17. In the preceding section we assumed that the sequences under consideration were finite. It is now desirable to drop this assumption and study preference orderings over sets of infinite sequences. Securities like the British *Consols* and the French *Rentes perpetuelles* offer an infinite sequence of interest payments, and an investor must have rules for comparing such securities with ordinary bonds, which mature after a finite number of years.

Koopmans [5] and [6] has studied preference orderings over infinite sequences of the form

$$x = \{x_1, x_2, \ldots, x_t, \ldots\}$$

and has shown that they can be represented by utility functions of a more general type than the classical discounting, i.e.,

$$u(x) = \sum_{t=1}^{\infty} v^t x_t.$$

A discussion of the nature of these utility functions would require fairly advanced mathematical methods, without giving the final solution to the problems which really interest us. We shall therefore only briefly indicate how uncertainty can be brought into the model.

12.18. A sequence of the type we have considered, $x = \{x_1, x_2, \ldots, x_t, \ldots\}$, where $x_1, \ldots, x_t, \ldots$ are stochastic variables, is called a discrete *stochastic process*. If x_t and x_s are stochastically independent for all $s \neq t$, the process is fairly easy to handle, and can be completely described by specifying the probability distributions of x_t for all t. If x_t and x_s are stochastically dependent, a description of the process must specify the nature of this dependence. General stochastic processes are difficult to handle, and there are satisfactory theories only for special cases, where the dependence between x_t and x_s is of special forms. As an illustration, let us consider a simple process which can be described as follows:

$$\text{Prob}(x_1 = 1) = \text{Prob}(x_1 = 0) = \tfrac{1}{2}.$$

If $x_t = 1$,

$$\text{Prob}\,(x_{t+1} = 1) = 1 - \alpha$$

$$\text{Prob}\,(x_{t+1} = 0) = \alpha.$$

If $x_t = 0$,

$$\text{Prob}\,(x_{t+1} = 1) = \alpha$$

$$\text{Prob}\,(x_{t+1} = 0) = 1 - \alpha.$$

A process of this kind is called a *Markov chain*, and is discussed in most textbooks of probability theory, e.g. [4]. The process can be described by two elements:

(i) An initial distribution, which in our example is $(\frac{1}{2}, \frac{1}{2})$.

(ii) A transition matrix, which in our example is

$$\begin{bmatrix} 1 - \alpha & \alpha \\ \alpha & 1 - \alpha \end{bmatrix}.$$

It is easy to verify that $E\{x_t\} = \frac{1}{2}$ for all t, so that the expected discounted sum of the sequence, i.e., the cash value, is

$$\sum_{t=1}^{\infty} v^t E\{x_t\} = \frac{v}{2(1 - v)}$$

for all values of α.

If the cash value is taken as the utility function which represents our preference ordering over a set of stochastic processes, all Markov chains obtained by letting α vary will be equivalent in this ordering. The resulting payment sequences are, however, very different, as we will see from the following examples:

(i) $\alpha = 0$: In this case we get either the sequence

$$1, 1, 1, 1, 1, \ldots$$

or

$$0, 0, 0, 0, 0, \ldots$$

each with probability $= \frac{1}{2}$.

(ii) $\alpha = \frac{1}{2}$: In this case we get a sequence where the terms are stochastically independent, and

$$\text{Prob}\,(x_t = 1) = \text{Prob}\,(x_t = 0) = \frac{1}{2} \quad \text{for all } t.$$

(iii) $\alpha = 1$: In this case we get an alternating sequence of the type

$$0, 1, 0, 1, 0, 1, 0, 1, \ldots$$

where the first term is determined by chance, being 0 or 1, each with probability $\frac{1}{2}$.

We may actually consider these three payment sequences as equally desirable, but we should nevertheless admit that there is a need for a theory in which such sequences are not treated as equivalent. The direction of preferences in a more general theory is by no means clear. Will the sequence become more desirable with increasing α, or do preferences go the other way?

12.19. A typical decision made by a firm, for instance a decision to make a new investment, will in general lead to a sequence of payments. There will usually be some uncertainty about the outcome of the investment, so that the resulting sequence must be considered as a stochastic process.

The problem of selecting the best among the available investments will then consist in selecting the best among the stochastic processes resulting from these investments. In order to solve this problem, the firm must have a *preference ordering over a set of stochastic processes*. The objective of the firm is then to obtain the profit sequence which corresponds to the most preferred of the attainable stochastic processes.

We have now somehow solved our problem and formulated the objective of the firm in a rational manner. It is, however, not easy to describe a preference ordering over a set of stochastic processes. It is therefore not surprising that businessmen have difficulties when they try to spell out the objectives pursued by their firms. The rules which we have discussed earlier may be considered as attempts to formulate a preference ordering which the decision-maker himself can see only in a very vague way.

REFERENCES

[1] Baumol, W. J.: *Economic Theory and Operations Analysis*, Prentice-Hall, 1961.
[2] Borch, K.: "Elements of a Theory of Reinsurance," *The Journal of Insurance*, September 1961, pp. 35–43.
[3] Dean, J.: *Managerial Economics*, Prentice-Hall, 1951.
[4] Feller, W.: *An Introduction to Probability Theory and its Applications*, Vol. 1, Wiley, 1950.
[5] Koopmans, T. C.: "Stationary Ordinal Utility and Impatience," *Econometrica*, 1960, pp. 287–309.
[6] Koopmans, T. C., P. A. Diamond, and R. E. Williamson: "Stationary Utility and Time Perspective," *Econometrica*, 1964, pp. 82–100.
[7] Wold, H.: "Ordinal Preferences or Cardinal Utility," *Econometrica*, 1952, pp. 661–663.

Chapter XIII

Survival and the Objectives
of the Firm

13.1. In Chapter XII we indicated how the objective of a firm could be formulated in such a general way that almost every objective conceived in the literature appears as a special case. However, the generality which made the model attractive for theoretical purposes may make it next to useless for practical application. It is hard to imagine a business consultant beginning his work by asking the company president to state his preferences over a set of stochastic processes. Should this happen, the president could, with some justification, reply that he does not know how such preferences should be expressed in a consistent manner, and he might also demonstrate that he knows very well how to express his opinion of certain consultants.

In this chapter we shall discuss a few applications of the general model. We shall not claim that these simple examples are realistic representations of economic situations in real life. Our main purpose is to show that a dynamic formulation of the problems may throw light on a number of the (essentially static) questions we have discussed in the preceding chapters.

13.2. We shall begin by considering a firm which operates under the following conditions:

(i) The firm has an initial capital S.
(ii) In each successive operating period the firm makes a profit x, which is a stochastic variable with distribution $F(x)$.
(iii) If the capital becomes negative at the end of an operating period, the firm is ruined and will have to go out of business.
(iv) If at the end of an operating period the capital exceeds Z, the excess will immediately be paid out as dividends.

In this model it is natural to take conditions (i) and (ii) as given. Condition (iii) is a "rule of the game" which can be relaxed, for instance by assuming that the firm can obtain credit when necessary. Condition (iv) represents a *dividend policy* of a very special kind, and it is natural to take Z as a decision variable which the firm can choose in the manner most suitable to its objective.

As a concrete interpretation we can think of our firm as an insurance company. Z will then be the reserves required for the kind of underwriting which the company expects to do in the foreseeable future.

13.3. It is easy to see that the model we have described will generate two discrete stochastic processes:

(i)
$$S_0, S_1, S_2, \ldots, S_t, \ldots$$

where S_t is the capital of the firm at the end of the tth operating period. It is obvious that $S_{t_0} < 0$ implies $S_t < 0$ for $t > t_0$.

(ii)
$$s_0, s_1, \ldots, s_t, \ldots$$

where s_t is the dividend paid at the end of period t.

The problem of the firm is now to determine the value of the only decision parameter Z so that the most preferred pair of the attainable stochastic processes is generated. Before this problem can be solved, the firm must lay down some rules as to when one attainable stochastic process is to be considered as better than another; i.e., the firm must formulate a preference ordering over a set of stochastic processes. In the following we shall discuss some fairly reasonable, and mathematically simple, orderings of this kind.

13.4. Let us first study process (i) in the preceding section, and consider the *expected number of non-negative* terms:

$$D(S, Z) = E\left\{ \sum_{t=0}^{\infty} \max\left(0, \frac{S_t}{|S_t|}\right) \right\}$$

where $S_0 = S$. $D(S, Z)$ is clearly the expected number of periods in which the firm will operate, or the "expected life" of the firm. If the objective of the firm is to survive as long as possible, it will prefer the process which gives the highest value to the function $D(S, Z)$; hence we obtain a preference ordering over the stochastic processes $S, S_1, \ldots, S_t, \ldots$ generated by our model.

From the conditions in 13.2 it follows that

$$D(S, Z) = 0 \qquad \text{for } S < 0$$

$$D(S, Z) = D(Z, Z) \quad \text{for } S > Z.$$

For $0 \le S \le Z$ it is easy to see that $D(S, Z)$ must satisfy the equation

$$D(S, Z) = 1 + \int_{-S}^{\infty} D(S + x, Z) \, dF(x).$$

When $S \ge 0$, the firm is certain to operate in at least one period. If the profit in this period is x, the firm can be expected to operate in $D(S + x, Z)$ additional periods. Multiplying by $dF(x)$, integrating over all x, and noting that $D(S + x, Z) = 0$ for $x < -S$, we obtain the equation.

If a density function $f(x) = F'(x)$ exists, the equation can be written

$$D(S, Z) = 1 + \{1 - F(Z - S)\}D(Z, Z) + \int_{0}^{Z} D(x, Z)f(x - S) \, dx.$$

Objectives of the Firm

This is an integral equation of Fredholm's type, with the simple kernel $f(x - S)$, and it can be solved by classical methods, which can be found in any textbook on integral equations. We can, for instance, form the iterated kernels

$$f^{(1)}(x - S) = f(x - S)$$
$$f^{(n)}(x - S) = \int_0^Z f^{(n-1)}(x - t)f(t - S) \, dt$$

and obtain the Liouville–Neumann expansion

$$D(S, Z) = 1 + \sum_{n=1}^{\infty} \int_0^Z f^{(n)}(x - S) \, dx$$

$$+ D(Z, Z)\left\{1 - F(S, Z) + \sum_{n=1}^{\infty} \int_0^Z \{1 - F(Z - x)\} f^{(n)}(x - S) \, dx\right\}$$

Here we can determine $D(Z, Z)$ by requiring the solution to be continuous at $S = Z$. This gives

$$D(Z, Z)\left\{F(0) - \sum_{n=1}^{\infty} \int_0^Z \{1 - F(Z - x)\} f^{(n)}(x - Z) \, dx\right\}$$

$$= 1 + \sum_{n=1}^{\infty} \int_0^Z f^{(n)}(x - Z) \, dx$$

13.5. Let us next study process (ii) in 13.3, and consider the *expected discounted* value of the dividend payments

$$V(S, Z) = E\left\{\sum_{t=0}^{\infty} v^t s_t\right\}$$

where $0 < v < 1$ is a discount factor. From the conditions in 13.2, it follows that

$$V(S, Z) = 0 \qquad \text{for } S < 0$$
$$V(S, Z) = S - Z + V(Z, Z) \quad \text{for } S > Z.$$

For $0 \le S \le Z$ the function $V(S, Z)$ must satisfy the equation

$$V(S, Z) = v \int_{-S}^{Z-S} V(S + x, Z) \, dF(x)$$

$$+ v \int_{Z-S}^{\infty} \{V(Z, Z) + x + S - Z\} \, dF(x).$$

If a density function $f(x) = F'(x)$ exists, the equation can be written

$$V(S, Z) = v \int_0^Z V(x, Z)f(x - S) \, dx$$

$$+ v \int_Z^{\infty} \{V(Z, Z) + x - Z\}f(x - S) \, dx$$

or

$$V(S, Z) = v \int_0^Z V(x, Z)f(x - S) \, dx$$

$$+ v\{1 - F(Z - S)\}V(Z, Z) + v \int_0^{\infty} xf(x + Z - S) \, dx.$$

[183]

This is again an integral equation of Fredholm's type, which can be solved by classical methods. It is clear that the function $V(S, Z)$ establishes an ordering over the set of stochastic processes $s_0, s_1, \ldots, s_t, \ldots$. This is the relevant preference ordering if the objective of the firm is to maximize the expected discounted value of the dividend payments which it will make during its lifetime.

The discount factor v does not necessarily have any connection with the market rate of interest. The factor expresses the "impatience" of the firm, i.e., the degree to which an earlier dividend payment is preferred to a later one.

13.6. The integral equations which we have derived in the two preceding sections can always be solved, but the solutions will usually be very complicated expressions which are next to impossible to discuss with elementary means. To bring out some of the properties of the solutions, we shall, therefore, discuss an extremely simple case. We shall assume that the distribution $F(x)$, introduced in 13.2, is defined as follows:

$$F(x) = 0 \quad \text{for} \quad x < -1$$

$$F(x) = q \quad \text{for} \quad -1 \le x < 1$$

$$F(x) = 1 \quad \text{for} \quad 1 \le x.$$

This means that in each operating period the firm can make

Either a gain of 1 with probability $p = 1 - q$,
or a loss of 1 with probability q.

The integral equation of 13.4 is then replaced by the difference equation

$$D(S, Z) = 1 + pD(S + 1, Z) + qD(S - 1, Z).$$

Similarly the integral equation in 13.5 is replaced by the difference equation

$$V(S, Z) = vpV(S + 1, Z) + vqV(S - 1, Z).$$

Both these equations can be solved by elementary methods. The first is really classical, and a solution can be found in the book by Feller ([2] p. 317). Both equations have been solved and discussed by de Finetti [3] in connection with a model very similar to the one considered in this chapter. The second equation has also been solved by Shubik and Thompson [7], who give several numerical examples, and by Borch [1].

These solutions are, however, valid only when S and Z are integers. We shall therefore solve the equations by a rather indirect method, which will make it easy to find $D(S, Z)$ and $V(S, Z)$ for arbitrary values of S and Z.

13.7. Let us assume that both S and Z are integers and let $w_n(S, Z)$ be the probability that the first dividend will be paid after n operating periods. It is easy to see that this probability must satisfy the recurrence relation

$$w_{n+1}(S, Z) = pw_n(S + 1, Z) + qw_n(S - 1, Z)$$

with the boundary conditions

$$w_0(S, Z) = 0 \quad 0 \le S \le Z$$
$$w_0(S, Z) = 1 \quad Z < S$$
$$w_n(S, Z) = 0 \quad S < 0$$
$$w_n(S, Z) = 0 \quad Z < S, n > 0.$$

The generating function

$$W(S, Z) = \sum_{n=0}^{\infty} v^n w_n(S, Z)$$

will give us the expected discounted value of the first unit paid as dividend. In our model the first payment will be equal to 1. Having made this payment, the firm will enter the next operating period with a capital Z; i.e., the expected discounted value of the later dividend payments will be $V(Z, Z)$.

From these considerations it follows that we have

$$V(S, Z) = \{1 + V(Z, Z)\}W(S, Z).$$

For $S = Z$ we find

$$V(Z, Z) = \frac{W(Z, Z)}{1 - W(Z, Z)}$$

Hence we have

$$V(S, Z) = \frac{W(S, Z)}{1 - W(Z, Z)}.$$

To determine the generating function, we note that $W(S, Z)$ must satisfy the difference equation

$$W(S, Z) = vpW(S + 1, Z) + vqW(S - 1, Z).$$

Let us first assume that this equation has a solution of the form

$$W(S, Z) = r^S.$$

Substitution in the original equation gives

$$r^S = vpr^{S+1} + vqr^{S-1}$$

or the *characteristic equation*

$$vpr^2 - r + vq = 0.$$

Hence the difference equation has a solution of the form r^s only if r is a root of the characteristic equation. This equation has the two roots

$$r_1 = \frac{1}{2vp}\{1 + (1 - 4v^2pq)^{1/2}\}$$

$$r_2 = \frac{1}{2vp}\{1 - (1 - 4v^2pq)^{1/2}\}.$$

It then follows that the difference equation is satisfied by an expression of the form

$$A_1 r_1{}^S + A_2 r_2{}^S$$

where A_1 and A_2 are arbitrary constants. One can show that this is the general solution of the difference equation.

From the boundary condition on $w_n(S, Z)$ at the beginning of this section it follows that we must have

$$W(-1, Z) = 0 \quad \text{and} \quad W(Z + 1, Z) = 1.$$

From these two equations we can determine the constants, and we find

$$A_1 = \frac{r_1}{r_1{}^{Z+2} - r_2{}^{Z+2}}, \qquad A_2 = \frac{-r_2}{r_1{}^{Z+2} - r_2{}^{Z+2}}$$

and hence

$$W(S, Z) = \frac{r_1{}^{S+1} - r_2{}^{S+1}}{r_1{}^{Z+2} - r_2{}^{Z+2}}$$

and further

$$V(S, Z) = \frac{(r_1{}^{S+1} - r_2{}^{S+1})}{(r_1{}^{Z+2} - r_2{}^{Z+2}) - (r_1{}^{Z+1} - r_2{}^{Z+1})}.$$

If we consider this as a continuous function of Z, it is easy to see that the denominator is minimized for a unique value of Z. It is tempting to assume that this value, which maximizes $V(S, Z)$, will give us an *optimal dividend* policy, at least within the class of policies we consider. We should, however, not jump to a conclusion, since the expressions we have found are valid only when S and Z are integers.

13.8. Let now S and Z be arbitrary numbers, and let $[S]$ be the greatest integer not exceeding S. We then make the following observations:

(i) If $Z - S$ is an integer, the first dividend will, if it is ever paid, be equal to 1.

   If $Z - S$ is not an integer, the first payment will be $1 - (Z - S) + [Z - S]$.

(ii) If $Z - S$ is an integer, the first payment can at the earliest be made after $(Z - S) + 1$ operating periods.

   If $Z - S$ is not an integer, the first payment can at the earliest be made after $[Z - S] + 1$ operating periods.

(iii) The capital of the company can become negative at the earliest after $[S] + 1$ operating periods.

From these observations it follows that we obtain $W(S, Z)$ for arbitrary non-negative values of S and Z from the expression in the preceding section if we replace

$$S \text{ by } [S]$$

and

$$Z \text{ by } [S] + [Z - S].$$

Hence we have

$$W(S, Z) = W([S], [S] + [Z - S])$$

and further

$$V(Z, Z) = V([Z], [Z])$$

and

$$V(S, Z) = \{1 - (Z - S) + [Z - S] + V(Z, Z)\} W(S, Z)$$

for $0 \le S \le Z$.

From this expression we see that $V(S, Z)$ is a piecewise linear function with jumps at the points where S, Z and $Z - S$ are integers.

We can find $D(S, Z)$ by similar methods. It is, however, easier to make a direct attack on the difference equation in 13.6:

$$D(S, Z) = 1 + pD(S + 1, Z) + qD(S - 1, Z).$$

The boundary conditions are

$$D(-1, Z) = 0$$

$$D(Z, Z) = D(Z + 1, Z).$$

From these we obtain the following expression, (see [2] p. 317) valid for integral values of, S and Z:

$$D(S, Z) = \frac{p}{(p - q)^2} \left\{ \left(\frac{p}{q}\right)^{Z+1} - \left(\frac{p}{q}\right)^{Z-S} \right\} - \frac{S + 1}{p - q}.$$

For arbitrary values $(0 \le S \le Z)$ we find

$$D(S, Z) = \frac{p}{(p - q)^2} \left\{ \left(\frac{p}{q}\right)^{[Z]+1} - \left(\frac{p}{q}\right)^{[Z-S]} \right\} - \frac{[S] + 1}{p - q}.$$

These expressions are meaningless for $p = q$. In the following we shall generally assume that $p > q$, i.e., that the sequence of gambles is favorable to the firm.

13.9. At this stage it may be useful to illustrate our findings by a simple numerical example. For this purpose we shall take $r_1 = 1.1$ and $r_2 = 0.7$. These values correspond approximately to

$$p = 0.565, \quad q = 0.435, \quad \text{and} \quad v = 0.983.$$

Our building stones are the functions $W(S, Z)$ and $V(S, Z)$ for integral values of S and Z. These are given in Tables 12 and 13 respectively.

TABLE 12
The generating function $W(S, Z)$

S	Z					
	0	1	2	3	4	5
0	0.56	0.40	0.33	0.28	0.24	0.20
1	1.00	0.73	0.59	0.50	0.43	0.39
2	—	1.00	0.80	0.69	0.60	0.53
3	—	—	1.00	0.85	0.75	0.66
4	—	—	—	1.00	0.87	0.75
5	—	—	—	—	1.00	0.89

TABLE 13
$V(S, Z)$ = *Expected discounted value of dividends*

S = Initial capital	Z = Capital required before dividends can be paid						
	0	1	2	3	4	5	6
0	1.25	1.49	1.70	1.83	1.89	1.88	1.82
1	2.25	2.69	3.05	3.30	3.41	3.40	3.27
2	3.25	3.69	4.19	4.52	4.67	4.65	4.49
3	4.25	4.69	5.19	5.56	5.79	5.79	5.56
4	5.25	5.69	6.19	6.56	6.81	6.80	6.55
5	6.25	6.69	7.19	7.56	7.81	7.80	7.50

We see that in this example the optimal dividend policy is given by $Z = 4$, at least as long as we only admit integral values of Z.

From these tables we can compute $V(S, Z)$ for non-integral values of the arguments. Some selected values of the function are given in Table 14. Table 15 gives the function $D(S, Z)$ for some selected values of S and Z.

TABLE 14
$V(S, Z)$ = *Expected discounted value of dividends*

S	Z									
	2.75	3.00	3.25	3.50	3.75	4.00	4.25	4.50	4.75	5.00
2.00	3.59	4.52	4.35	4.17	4.00	4.67	4.53	4.37	4.20	4.65
2.25	3.79	4.72	4.52	4.35	4.17	4.85	4.67	4.53	4.37	4.82
2.50	3.99	4.92	4.72	4.52	4.35	5.02	4.85	4.67	4.53	4.97
2.75	4.19	5.12	4.92	4.72	4.52	5.19	5.02	4.85	4.67	5.12
3.00	4.44	5.56	5.12	4.92	4.72	5.79	5.19	5.02	4.85	5.79
3.25	4.69	5.81	5.56	5.12	4.92	6.01	5.79	5.19	5.02	5.97
3.50	4.94	6.06	5.81	5.56	5.12	6.22	6.01	5.79	5.19	6.16
3.75	5.19	6.31	6.06	5.81	5.56	6.43	6.22	6.01	5.79	6.35
4.00	5.44	6.56	6.31	6.06	5.81	6.81	6.43	6.22	6.01	6.80
4.25	5.69	6.81	6.56	6.31	6.06	7.07	6.81	6.43	6.22	7.03

TABLE 15
$D(S, Z) = $ *Expected Life of the Firm*

S	Z						
	0	1	2	3	4	5	6
0	2.3	5.3	9.2	14.3	20.9	29.5	40.7
1	2.3	7.6	14.6	23.5	35.2	50.2	69.9
2	2.3	7.6	16.9	29.0	44.4	64.7	79.1
3	2.3	7.6	16.9	31.2	50.0	74.0	105.4
4	2.3	7.6	16.9	31.2	52.3	79.3	115.1
5	2.3	7.6	16.9	31.2	52.3	81.5	119.9
6	2.3	7.6	16.9	31.2	52.3	81.5	123.2

13.10. We shall now discuss our model and indicate how it is relevant to a number of questions discussed in the preceding chapters. Let us first assume that the objective of our firm is to maximize the expected discounted value of the dividends which will be paid during its lifetime. From the shareholders' point of view this appears to be a desirable objective, and it clearly leads to the problem of determining the value of Z which maximizes $V(S, Z)$ for a given S.

Let us take $S = 3.5$, interpreting this as the actual working capital. Let us further make the (possibly unrealistic) assumption that the Board of Directors has accepted the objective desired by the shareholders. We can then imagine that the Board debates whether a dividend should be paid or not. It may decide that the present capital constitutes a reserve which is just sufficient, i.e., that only if the capital should increase above $S = 3.5$ will the excess be paid out as dividend. From Table 14 we see that this policy decision means that the expected discounted value of the firm's dividend payments will be

$$V(3.5, 3.5) = 5.56.$$

The Board may then consider some *small* changes in this policy. It may, for instance, propose to pay a dividend of 0.25, and continue operating with a reserve capital $Z = 3.25$. This will increase the expectations to

$$V(3.5, 3.25) = 0.25 + V(3.25, 3.25) = 5.81.$$

The Board may also consider postponing dividend payment and setting the reserve requirements at $Z = 3.75$. This will reduce dividend expectations to

$$V(3.50, 3.75) = 5.12.$$

It is easy to see from Table 14 that if the Board only considers reserve requirements in the neighborhood of the present capital $S = 3.5$, it may well arrive at the conclusion that the optimal reserve capital is $Z = 3$, giving the dividend expectations

$$V(3.5, 3.0) = 0.5 + V(3.3) = 6.06.$$

If, however, the Board is prepared to consider more ambitious reserve schemes, it should be able to discover that the real optimum may be $Z = 4$, with the corresponding expectations

$$V(3.5, 4) = 6.22.$$

13.11. This simple case provides an example of how a firm can be led to make the "wrong" decision by using techniques of analysis which give the nearest *local optimum*. Such mistakes are probably made in business, particularly by economists who are fond of calculus and marginal analysis. We shall therefore study how the situation may develop after an initial mistake of this kind.

In our example the firm will pay a dividend of 0.5, and continue operating with the declared policy that whenever the capital exceeds 3 the excess is to be paid out as dividends. If the firm is not ruined, its capital will sooner or later increase to $S = 4$. The firm may then stick to its declared policy and pay a dividend of 1. The firm may, however, discover that the local optimum technique indicates that it will pay to change the policy. We see from Table 14 what the dividend expectations are:

(i) By adhering to the established policy

$$V(4, 3) = 1 + V(3.3) = 6.56.$$

(ii) By changing reserve requirements to $Z = 4$, the expectation becomes

$$V(4, 4) = 6.81.$$

This is, as we see from Table 14, the optimal policy. When we observe actual business behavior, we may well find processes which by some stretch of the imagination resemble the one we have outlined. Economists, aided by psychologists, may then think up strange theories to explain these observations. They may, for instance, suggest that firms tend to become more greedy as they become richer; i.e., the more capital they accumulate, the more capital will they retain as "necessary" reserves. Such exotic theories may contain a grain of truth, but are they really necessary? The simplest explanation is that the firms we observe are not able to handle the mathematics of their decision problem.

If the firm solves the problem properly, it should be clear that the situation is as illustrated in Fig. 14. It may then be possible for the Board to convince the impatient shareholders that it is in their long-term interest that dividend payments be postponed.

13.12. Let us now assume that our firm has solved the problem in the preceding section, and has found the value of Z which corresponds to the

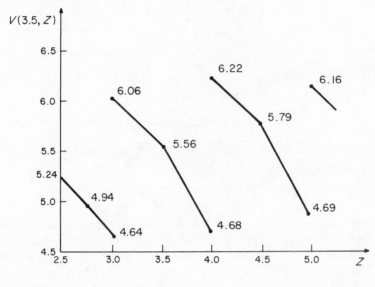

Figure 14

optimal dividend policy. Let us further assume that at this point the firm is offered a prospect of the type

Gain R_1 with probability α
Lose R_2 with probability $1 - \alpha$.

We can think of this as a short-term investment. If our firm accepts the offer, the expected value of its future dividend payments will become

$$\alpha V(S + R_1, Z) + (1 - \alpha)V(S - R_2, Z).$$

If the overall objective of maximizing the expected discounted value of dividend payments is maintained, the offer will be accepted if and only if

$$\alpha V(S + R_1, Z) + (1 - \alpha)V(S - R_2, Z) > V(S, Z).$$

This means, however, that the decision will be made as if the firm wanted to maximize expected utility, with $V(S, Z)$ serving as the utility function.

Let us now return to the expression from 13.7

$$V(S, Z) = \frac{(r_1^{S+1} - r_2^{S+1})}{(r_1^{Z+2} - r_2^{Z+2}) - (r_1^{Z+1} - r_2^{Z+1})}$$

and look at it as a continuous function in S and Z. For simplicity we shall write

$$V(S, Z) = \frac{N(S)}{M(Z)}.$$

[191   ]

The optimal dividend policy Z_0 is obviously determined by the condition
$$M'(Z_0) = 0$$
or
$$\left(\frac{r_1}{r_2}\right)^{Z_0+1} = \frac{r_2 - 1}{r_1 - 1}\frac{\log r_2}{\log r_1}.$$
It is easy to verify that
$$N'(S) > 0$$
and that
$$N''(S) < 0 \quad \text{for } 0 < S < S_0$$
$$N''(S) > 0 \quad \text{for } S_0 < S$$
where S_0 is determined by
$$\left(\frac{r_1}{r_2}\right)^{S_0+1} = \left(\frac{\log r_2}{\log r_1}\right)^2.$$

From these expressions we see that S_0 and Z_0 are approximately equal. This means that $V(S, Z)$ as a function of S has an inflexion point near $S = Z_0$. Recalling that $V(S, Z)$ is linear in S for $Z < S$, we can pull the bits together:

(i) If reserve requirements are set lower than optimal, i.e., $Z < Z_0$, $V(S, Z)$ as a function of S will be concave for $0 \leq S \leq Z$ and linear for $Z < S$.

(ii) If reserve requirements are set higher than optimal, i.e., $Z_0 < Z$, $V(S, Z)$ as a function of S will be

$$\text{Concave} \quad \text{for } 0 \leq S \leq S_0$$
$$\text{Convex} \quad \text{for } S_0 < S \leq Z$$
$$\text{Linear} \quad \text{for } Z < S.$$

In the latter case, $V(S, Z)$ as utility function can lead to the paradoxical behavior observed by Friedman and Savage [4]. However, to select a policy corresponding to $Z > Z_0$ is really inconsistent with our assumption that the firm's objective is to maximize the expected discounted value of its dividend payments. Hence it appears that the observations of Friedman and Savage may be explained by contradictions between the short-term and long-term behavior of the decision-maker.

Another explanation may be that the firm may for some reason be forced to keep larger reserves than it considers optimal. Some restrictions of this kind may well be imposed on banks and insurance firms by government regulations.

It appears in general that $V(S, Z)$, interpreted as a utility function, has the properties required to explain most of the economic behavior which we can observe.

For $S < 0$ the function $V(S, Z)$ is constant, equal to zero. This is quite natural in our simple model, where the worst thing which can happen is that our firm is forced out of business. By introducing "fates worse than death" we can obviously obtain utility functions of more orthodox shape.

The least desirable property of our utility function may be that it is linear for $S > Z$. This of course is due to the simplicity of our model, and we may interpret it to mean that a decision rule which works for small everyday problems breaks down when applied to large amounts of money.

We can also construct an explanation to the effect that our firm, seeing no immediate business use for amounts greater than Z, automatically uses a linear utility in decisions where such amounts are involved.

13.13. So far we have found that the expression for $V(S, Z)$ appears quite reasonable as a utility function if we look at it as a continuous function. However, $V(S, Z)$ is only piecewise continuous, as indicated by Fig. 15, and this can obviously lead our firm to make decisions which will seem odd, if not outright silly, to an outsider, for instance to an economist who wants to study business behavior.

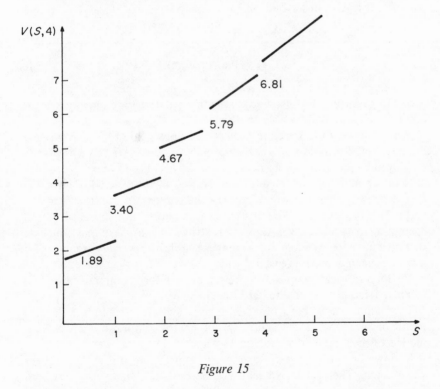

Figure 15

As an illustration let us assume that the firm's capital is $S = 2.75$ and that it has adopted the optimal dividend policy, corresponding to $Z = 4$.

From the corresponding column in Table 14 we see that the expected discounted value of the future dividend payments is

$$V(2.75, 4) = 5.19.$$

To make the discussion concrete, let us assume that the firm is an insurance company and that this company is offered a premium of 0.5 if it will accept an insurance contract which can lead to a claim payment of 1 with probability 0.6. The proposed contract is obviously unfavorable to the company, and any orthodox actuary would reject it summarily.

If, however, the company, in spite of this, accepts the contract, the expected discounted value of future dividend payments will be

$$0.4V(3.25, 4) + 0.6V(2.25, 4) = 5.31.$$

Hence the company should accept this unfavorable contract if it is consistent in pursuing its long-term objective of maximizing the expected discounted value of its dividend payments.

If, however, the company's capital is 3.25, it should not accept this contract. It is easy to see that we have

$$V(3.25, 4) = 6.01$$

and

$$0.4V(3.75, 4) + 0.6V(2.75, 4) = 5.69;$$

i.e., acceptance will lead to a reduction of the expected value of future dividend payments. We also find that in this situation the company will not accept a reasonably favorable contract.

If an outsider observes a number of decisions made by this company, he may well conclude that the management is crazy. If this observer is employed by the company as a consultant, he may have to take the rationality of top management as an axiom. He will then be forced to conclude that the utility function which governs the company's decisions wiggles, and that it may be approximated by a combination of a straight line and something like a sine curve. As an alternative, our consultant may conclude that rational people simply do not make their decisions in accordance with the Von Neumann–Morgenstern utility theory. He is then forced to conclude that rational people must reject one of the axioms discussed in Chapter III, and this may be hard to explain.

The paradoxes we have observed are evidently caused by the discrete nature of our model, and we may expect them to disappear in more general models of the type we discussed in 13.4 and 13.5.

If, however, our firm really operates in a discrete world, the paradoxes will remain. The real world may not be discrete, but a firm may well formalize its expectations about future business in a discrete manner. If the firm acts consistently on these expectations, it will have to accept unfavorable prospects in some situations.

13.14. Let us now take a different approach and assume that the main concern of the firm is to *survive* as long as possible. The objective of the firm may then be to maximize the function $D(S, Z)$. From the expression in 13.8,

$$D(S, Z) = \frac{p}{(p - q)^2} \left\{ \left(\frac{p}{q}\right)^{Z+1} - \left(\frac{p}{q}\right)^{Z-S} \right\} - \frac{S + 1}{p - q},$$

it is clear that $D(S, Z)$ goes to infinity with Z. This means that the optimal dividend policy is not to pay any dividend at all, no matter how large the capital of the firm becomes. Hence the model is trivial unless we assume that some upper limit is imposed on Z. For some firms this is a natural assumption. A mutual insurance company may, for instance, be obliged to pay back to the policyholders any capital which cannot be considered as "necessary reserves" for the company's operations.

Let us assume that the upper limit on the capital which the company can accumulate is $\bar{Z}$, and that the actual capital is $S \leq \bar{Z}$. If the company receives an unexpected offer of the type described in 13.12, it will accept the offer if and only if

$$\alpha D(S + R_1, \bar{Z}) + (1 - \alpha)D(S - R_2, \bar{Z}) > D(S, \bar{Z}),$$

i.e., if the company can increase its expected life by accepting the offer.

From Table 15, and from the expression for $D(S, Z)$ taken as a continuous function, it is clear that $D(S, \bar{Z})$ has most of the properties desirable in a conventional utility function:

(i) $D(S, \bar{Z})$ is an increasing function of S.
(ii) $D(S, \bar{Z})$ is bounded from above, because $D(S, \bar{Z}) = D(\bar{Z}, \bar{Z})$ for $S > \bar{Z}$.
(iii) Marginal utility is decreasing.

Hence, if the objective of the firm is to maximize its expected life, it will show a strong risk aversion, and in general behave in good accordance with current economic theory. There will be no paradoxes of the kind observed by Friedman and Savage. $D(S, Z)$ is, of course, really a discontinuous function, and this may lead to behavior which may be considered paradoxical.

13.15. It seems natural to assume that the owners of the firm, i.e., the shareholders, will want the firm to operate so that $V(S, Z)$ is maximized. The daily management of the firm may, however, be in the hands of an executive who in reality is the paid employee of the owners. If this executive is absolutely loyal to his employers, he should in every situation make the decision which leads to the highest possible value of $V(S, Z)$. But it is not unreasonable to assume that the executive has objectives of his own which do not necessarily coincide with the objectives of the owners. If the main concern of the executive is to perpetuate his own position, he will probably, whenever possible, make decisions which tend to maximize $D(S, Z)$.

The possibility of such conflicts of interest between owners and the executives who make the decisions on behalf of the firm has been observed by most economists who have studied the objectives of firms. It is recognized, that such conflicts of interest may lead a firm to behave in an erratic manner, depending on the set of objectives which dominates any particular decision. This is, of course, well known in business circles. where a dividend is defined as a profit which can no longer be hidden from the shareholders.

It seems that our simple model provides some elements which may help us to understand and analyze such conflicts of interest within the firm, and thus pave the way toward a more realistic economic theory.

13.16. Our model may also throw some light on the theoretical problem of assigning a specific utility to gambling. To illustrate this, let us assume that our firm has a capital $S = 5$ at the end of an operating period and that the Board of Directors debates whether a dividend should be paid or not, i.e., how high Z (the required reserve) should be set.

From Table 13 the Board will see that dividend expectations are maximized for $Z = 4$, i.e., if a dividend of 1 is paid.

Consulting Table 15, the Board will see that this action will give

Expected life: 52.3
Dividend expectations: $1 + 6.81 = 7.81$.

Similarly the decision of setting reserve requirements at $Z = 6$ will give

Expected life: 119.9
Dividend expectations: 7.55.

Would we be surprised if the Board in this situation prefers the second decision?

Our question may contain a trap, as did the question of Allais which we discussed in Chapter VI. If a person, or a board, selects $Z = 6$ because it gives a much longer expected life, the reason may be that he subconsiously believes that a long life in business means a long sequence of dividend payments, with a high expected discounted value. We may point out to our person that this belief is false. Then he may, or he may not, change his mind and select $Z = 4$.

13.17. If the decision-maker, or our Board, insists that his choice is $Z = 6$, he must attach some specific utility to merely staying in business which can compensate for a reduction of the monetary payoff. In more general terms this means that there is some "specific utility of gambling," apart from the possibilities of gain offered by the gambling.

This "specific utility of gambling" cannot be accommodated in the utility theory of Von Neumann and Morgenstern, who remark, "This may seem to be a paradoxical assertion. But anybody who has seriously tried to axiomatize that elusive concept will probably concur with it" ([6] p. 28). Returning at the end of their book to the question, they state, "It constitutes a much deeper problem to formulate a system, in which gambling has under all conditions a definite utility or disutility," ([6] p. 629) and "Some change of the system [i.e. of the axioms] may perhaps lead to a mathematically complete and satisfactory calculus of utilities, which allows for the possibility of a specific utility or disutility of gambling. It is hoped that a way will be found to achieve this, but the mathematical difficulties seem to be considerable. Of course, this makes the fulfillment of the hope of a successful approach by purely verbal means appear even more remote" ([6] p. 632).

Our discussion of an extremely simple model obviously does not provide the theory desired by Von Neumann and Morgenstern, but it may point out the way which eventually will lead to this theory. In our model we have not run into any formidable mathematical problems. However, in order to come to grips with the "elusive concept" of Von Neumann and Morgenstern, we had to take their essentially static decision problem and see it in its natural dynamic context, i.e., in an infinite sequence of such decision problems, and this means that we have in a sense placed the whole discussion on a different level.

13.18. So far we have taken ourselves and our model very seriously and have tried to relate it to the decisions which are made in the Board Rooms, and which are assumed to determine the destiny of the world. However, we can give different interpretations to our model; in a lighter vein, we can call it *the drinking man's game.*

Let us consider a gambler who enters a casino with a capital S and assume

(i) He enjoys gambling for its own sake, so that he wants his capital to last as long as possible.

(ii) He also enjoys having a drink at his side when gambling, and he wants his drink as early as possible.

The problem of this gambler is to find a policy for ordering drinks which is optimal in the sense that it strikes the best possible balance between his desires for gambling and drinking.

It is clear that if he is really obsessed with gambling, he will never waste money on drinks, since this will shorten the time he can expect to stay in the casino. Similarly if the main concern of our gambler is drinking, he will order as many drinks as he can, keep just a minimum stake, and play with this as long as his luck lasts. If he is a normal person, with no extreme tastes, he may strike the proper balance by determining the value of Z which maximizes a utility function $U\{V(S, Z), D(S, Z)\}$, and order a drink whenever his capital exceeds Z.

13.19. In our discussion of a simple example, we made some surprising findings, which obviously were due to the discrete nature of the basic stochastic variable. It may, therefore, be useful to give the corresponding results for a continuous example, even if the mathematics required for a full analysis is not strictly elementary. Let us assume

$$f(x) = k\alpha e^{-\alpha x} \qquad \text{for } x > 0$$
$$f(x) = (1 - k)\alpha e^{\alpha x} \quad \text{for } x < 0.$$

We shall assume that $\frac{1}{2} < k < 1$, i.e., that the business is favorable to the firm. The integral equation from 13.5 can now be written as follows:

$$V(S) = v(1 - k)\alpha e^{-\alpha S} \int_0^S V(x)e^{\alpha x}\, dx$$

$$+ vk\alpha e^{\alpha S} \int_S^Z V(x)e^{-\alpha x}\, dx$$

$$+ vkV(Z)e^{\alpha(S - Z)} + \frac{vk}{\alpha} e^{\alpha(S - Z)}.$$

For simplicity we have written $V(S)$ for $V(S, Z)$, since there should be no risk of misunderstanding.

Differentiating the integral equation twice with respect to S, we see that $V(S)$ must satisfy the differential equation

$$(1 - v)\alpha^2 V(S) + v(1 - 2k)\alpha V'(S) - V''(S) = 0.$$

The general solution of this equation is

$$V(S) = C_1 e^{r_1 S} + C_2 e^{r_2 S}.$$

Here C_1 and C_2 are arbitrary constants, and r_1 and r_2 are the roots of the characteristic equation

$$r^2 - v(1 - 2k)\alpha r - (1 - v)\alpha^2 = 0.$$

It is easy to verify that these roots both are real, and that $r_1 > 0$, $r_2 < 0$.

The constants C_1 and C_2 must be determined so that the general solution of the differential equation also is a solution of the integral equation. Substituting the general solution in the integral equation, we find

$$C_1 = \frac{-1}{\alpha(r_2 + \alpha)M}, \qquad C_2 = \frac{1}{\alpha(r_1 + \alpha)M}$$

where

$$M = \frac{r_1 e^{r_1 Z}}{(r_1 - \alpha)(r_2 + \alpha)} - \frac{r_2 e^{r_2 Z}}{(r_1 + \alpha)(r_2 - \alpha)}.$$

This gives us the following explicit expression for the expected discounted value of the dividend payments:

$$V(S, Z) = \frac{1}{\alpha M} \left\{ \frac{e^{r_2 S}}{r_1 + \alpha} - \frac{e^{r_1 S}}{r_2 + \alpha} \right\}.$$

By similar consideration we find that $D(S) = D(S, Z)$ must satisfy the integral equation

$$D(S) = 1 + (1 - k)\alpha e^{-\alpha S} \int_0^S D(x)e^{\alpha x} \, dx$$

$$+ k\alpha e^{\alpha S} \int_S^Z D(x)e^{-\alpha x} \, dx + ke^{\alpha(S - Z)} D(Z).$$

Differentiating twice, we find that the integral equation can be reduced to the differential equation

$$(2k - 1)\alpha D'(S) + D''(S) + \alpha^2 = 0.$$

The general solution of this equation is

$$D(S) = C_1 e^{-(2k - 1)\alpha S} - \frac{\alpha}{2k - 1} S + C_2$$

where C_1 and C_2 are constants which must be determined so that the solution also satisfies the integral equation.

As a numerical example, let us take $\alpha = 1$, $r_1 = 0.1$, and $r_2 = -0.3$. This corresponds to $v = 0.97$ and $k = 0.603$. We then find

$$V(S, Z) = \frac{143e^{0.1S} - 91e^{-0.3S}}{16e^{0.1Z} + 21e^{-0.3Z}}$$

and

$$D(S, Z) = 37.5e^{0.2Z} - 5(1 + S) - 30e^{0.2(Z - S)}.$$

Table 16 gives the value of the function $V(S, Z)$ for some selected values of S and Z. It is easy to verify that this function takes its maximal value of $Z = 3.45$. Table 16 gives the value of the function $D(S, Z)$ for the same values of S and Z.

TABLE 16
$V(S, Z) = $ *Expected discounted value of dividend payments*

| | | | Z | | | |
S	0	1	2	3	4	5
0	1.41	1.57	1.68	1.74	1.73	1.68
1	2.41	2.74	2.93	3.02	3.02	2.94
2	3.41	3.74	4.03	4.16	4.16	4.04
3	4.41	4.74	5.03	5.21	5.20	5.14
4	5.41	5.74	6.03	6.21	6.19	6.02
5	6.41	6.74	7.03	7.21	7.19	6.98

TABLE 17
D(S, Z) = Expected life of the firm

S	Z					
	0	1	2	3	4	5
0	2.5	4.2	6.2	8.7	11.7	15.4
1	2.5	5.8	9.6	13.3	19.0	25.1
2	2.5	5.8	11.2	16.7	23.6	32.4
3	2.5	5.8	11.2	18.3	27.0	37.0
4	2.5	5.8	11.2	18.3	28.6	40.4
5	2.5	5.8	11.2	18.3	28.6	42.0

13.20. It may be useful to conclude this chapter by placing our findings in their historical context. The original purpose of probability theory was to find rules for selecting the best from a set of available prospects. The first rule offered as a solution was to select the prospect which has the greatest *expected gain*.

Daniel Bernoulli observed that a rich man and a poor man would not— and should not—apply the same decision rule. He then concluded that the correct rule would have to depend on the *wealth* of the decision-maker, and his rule was to select the prospect with the greatest *expected utility*.

In this chapter we have argued that "wealth" must be taken not just as cash in hand. The relevant concept must include all future prospects of the decision-maker, i.e., the gambles which he expects to take in his lifetime. This is in some ways in good accordance with classical economic theory, which assumes that the utility of money must be derived from our expectations of the future use we can make of money.

The mathematical model we have discussed is usually referred to as a *random walk* with one absorbing and one reflecting barrier. For discrete probability distributions the model can be analyzed by fairly elementary mathematical techniques, and it is discussed in most textbooks of probability theory, for instance by Feller ([2] Chap. XIV). For continuous probability distributions the model leads to fairly complicated mathematical problems. These have been discussed in considerable detail in a recent book by Kemperman [5], which also contains a comprehensive bibliography.

REFERENCES

[1] Borch, K.: "The Optimal Management Policy of an Insurance Company," *Proceedings of the Casualty Actuarial Society*, Vol. LI (1964), pp. 182–197.
[2] Feller, W.: *An Introduction to Probability Theory and its Applications*, 2nd ed., Wiley, 1957.
[3] Finetti, B. de: "Su una Impostazione Alternativa della Teoria Collettiva del Rischio," *Transactions of the XV International Congress of Actuaries*, 1957, New York, Vol. II, pp. 433–443.

[4] Friedman, M. and L. J. Savage: "The Utility Analysis of Choices Involving Risk," *Journal of Political Economy*, 1948, pp. 279–304.

[5] Kemperman, J. H. B.: *The Passage Problem for a Stationary Markov Chain*, University of Chicago Press, 1961.

[6] Neumann, J. von and O. Morgenstern: *Theory of Games and Economic Behavior*, 2nd ed., Princeton University Press, 1947.

[7] Shubik, M. and G. Thompson: "Games of Economic Survival," *Naval Research Logistics Quarterly*, 1959, pp. 111–123.

Chapter XIV

Credibility and Subjective Probabilities

14.1. In the preceding chapters we have generally assumed that the decision-maker knows the probabilities with which the different states of the world will occur. In real life, a decision-maker will often argue that he does not really know these probabilities. If we then ask him if he is so completely ignorant that he could use the Laplace Principle of Insufficient Reason and assign equal probabilities to all states of the world, it is likely that he will reply that he is not quite as ignorant as that. He feels that he knows something about the possible states of the world, but he finds it hard to spell out exactly what he knows or believes he knows.

In Chapter VII we tried to bypass this problem by formulating decision rules which were independent of the probabilities of the different states of the world. In this chapter we shall study the ways in which vague knowledge or beliefs about these probabilities can be brought to bear on decision problems. The ideas which we shall develop were first formulated explicitly, and in a general manner, by Savage [5].

To make our discussion concrete, we shall consider an insurance company which holds a capital S and a portfolio of insurance contracts. We shall assume that the total amount of claims which will be paid under the contracts in the portfolio is a stochastic variable x with the distribution $G(x)$. We shall assume that this distribution is known to the company. The company will assign the following utility to this situation:

$$U(S) = \int_0^\infty u(S - x) \, dG(x)$$

where $u(x)$ is the utility function which represents the company's preference ordering, or attitude to risk. Since we have not brought the time element into the model, we shall assume that all contracts in the portfolio expire within a fairly short period.

14.2. Let us now assume that the company is offered an additional contract which will expire within the same short period. Let $P =$ the premium paid for this contract, and let the claim distribution be

Z with probability p
0 with probability $1 - p = q$.

The company will accept the new contract if and only if it leads to an increase in utility, i.e., if

$$pU(S + P - Z) + qU(S + P) > U(S).$$

In some contexts it may be more natural to assume that the company is invited to "quote a premium" for the new contract. The equation

$$pU(S + P - Z) + qU(S + P) = U(S)$$

will then determine the lowest premium P which the company can quote. If the elements $u(x)$ and $G(x)$ are known, we can find this premium by simple application of the principles we have developed in earlier chapters. We can then study how the company should bargain to obtain a higher premium.

In a real-life situation the company may not feel so certain that the relevant probability is exactly p. Our problem is to find out what this actually may mean, and to study how the insurance company will, or should, make its decision in such situations.

Let us first assume that the insurance company maintains that it *knows absolutely nothing* about the risk to be covered by the new contract. If this statement has any meaning at all, it must imply that any value of p between 0 and 1 is equally possible—or equally probable. It is then natural to write the equation from the preceding paragraph in the following form

$$p\{U(S + P - Z) - U(S + P)\} + U(S + P) = U(S),$$

multiply by dp, and integrate from 0 to 1. This will give

$$\tfrac{1}{2}\{U(S + P - Z) + U(S + P)\} = U(S)$$

as the equation which determines the lowest premium P which the company should quote. This implies that the company should act as if $p = \tfrac{1}{2}$.

14.3. In practice we will not often have to make decisions under complete ignorance. We will usually have some information or *prior belief* about p. The mere fact that somebody wants to pay for this insurance contract indicates that p is not zero; i.e., the event which will lead to a claim payment is not impossible.

The usual procedure may be as follows: The actuary and the more or less experienced underwriters may agree that the "best estimate" is, say, $p = 0.10$, adding that this is little more than an educated guess. When pressed for more precision, they may state that p is very likely to be somewhere in the interval $(0.05, 0.20)$, or that they are certain that $p < 0.40$.

Arguments of this kind reflect a vague, but very real, feeling of uncertainty. The most natural way of giving precision to this statement seems to be to specify the weights which should be given to the various possible values of p. We can do this by specifying a function $f(p)$ which takes its greatest value for the "best estimate" or "most likely" value of p.

There is obviously nothing to prevent us from normalizing the function, and requiring that

$$\int_0^1 f(p)\, dp = 1.$$

This means that the function $f(p)$, which represents our belief about the value of p, has all the properties of a probability distribution.

So far this makes sense. The trouble comes if or when we state something like the following:

$f(0.1) = \text{Prob}\{p = 0.1\} = \text{the probability that the parameter } p \text{ is equal to } 0.1.$

This statement has no real meaning. A parameter is not a stochastic variable, so it has no meaning to assign a probability other than 0 and 1 to the "event" that it takes a particular value.

14.4. If our insurance company can specify the function which represents its prior belief, the decision problem is solved by multiplying the basic equation by $f(p)\, dp$ and integrating from 0 to 1. This gives

$$U(S + P) + \{U(S + P - Z) - U(S + P)\} \int_0^1 pf(p)\, dp = U(S)$$

or, if we write

$$\bar{p} = \int_0^1 pf(p)\, dp,$$

$$\bar{p}U(S + P - Z) + (1 - \bar{p})U(S + P) = U(S).$$

This means that the company acts *as if* it was certain that the parameter has the value $\bar{p}$, i.e., the "expected value" of the prior belief.

To obtain a more realistic example, let us assume that the company is offered a portfolio of n contracts of the type considered, so that the amount of premium received is nP. If k of the contracts should lead to a claim payment, the company will have to pay a total amount $y = kZ$. The probability of this event is

$$\text{Prob}\{y = kZ\} = \binom{n}{k}p^k(1 - p)^{n-k}$$

provided that claim payments under different contracts are stochastically independent. The minimum premium is then determined by the equation

$$U(S) = \sum_{k=0}^n U(S + nP - kZ)\binom{n}{k}p^k(1 - p)^{n-k}.$$

This formula takes into account the uncertainty which, in statistical language, is due to "sampling fluctuations." The number of payments, k, can take any value from 0 to n. We know the probabilities of these $n + 1$ events, and we use these probabilities to compute the weighted sum of the utility corresponding to each event.

The uncertainty due to incomplete knowledge about the true value of the parameter p is of a different logical nature. If, however, we are willing to specify a *subjective probability* distribution $f(p)$, which represents our "prior belief," we can deal also with this second kind of uncertainty.

Multiplying the basic equation by $f(p) \, dp$, and integrating from 0 to 1, we find that the minimum premium is determined by

$$U(S) = \sum_{k=0}^{n} U(S + nP - kZ) \left\{ \binom{n}{k} \int_0^1 p^k (1 - p)^{n-k} f(p) \, dp \right\}$$

or

$$U(S) = \sum_{k=0}^{n} U(S + nP - kZ) h(k)$$

if we write

$$h(k) = \binom{n}{k} \int_0^1 p^k (1 - p)^{n-k} f(p) \, dp.$$

From this equation we see that the second kind of uncertainty does not change the mathematical structure of the problem. The minimum premium is determined by an equation of the same form as earlier, but the binomial distribution has been replaced by the distribution $h(k)$, $(k = 0, 1, \ldots, n)$.

14.5. Let us now assume that the insurance company of the preceding section really knows nothing about the probability that a contract will lead to a claim. There may then be something to be said for giving all values of p the same weight, i.e., taking $f(p) = 1$, and making the decision on this basis. We then find

$$h(k) = \binom{n}{k} \int_0^1 p^k (1 - p)^{n-k} \, dp = \frac{1}{n+1};$$

i.e., the company will make its decision as if all values of k were equally probable.

In practice it is not likely that we will be so completely ignorant that this procedure is justified. Usually we will have some relevant information about the probabilities. For an insurance company it is natural to assume that such information consists of the knowledge that in a portfolio of n comparable contracts there were k claim payments. This knowledge may have been derived from the company's experience of last year's underwriting, or from published statistics.

To a statistician, it is then natural to suggest that the company should act as if

$$p = k/n.$$

He will usually be able to justify this in several different ways. There is, however, some uncertainty about this estimate, particularly if n is small. The statistician will probably be able to give a number which measures this uncertainty, but he may be unable to explain how the company should allow for this uncertainty when making its decision.

Let us approach this problem by writing

$$\text{Prob}(k \mid p) = \binom{n}{k} p^k (1 - p)^{n-k}.$$

This is the probability of the observed result k if the true probability is p. This is usually referred to as the *likelihood* of the observed result. One justification for taking $p = k/n$ is that this value will maximize the likelihood.

From the theory of conditional probability we know that

$$\text{Prob}(k \mid p)\,\text{Prob}(p) = \text{Prob}(p \mid k)\,\text{Prob}(k)$$

or

$$\text{Prob}(p \mid k) = \frac{\text{Prob}(k \mid p)\,\text{Prob}(p)}{\text{Prob}(k)}.$$

Here the denominator can be interpreted as the absolute probability of k, i.e., the probability of observing k claims regardless of what the true probability p may be.

We then have in a purely formal way

$$\text{Prob}(k) = \sum_p \text{Prob}(k \mid p)\,\text{Prob}(p).$$

Here the sum is over all values of p, and Prob (p) is the weight of our belief that the true value of the parameter is p. To express this result in the notation we have used earlier, we shall write

$$\text{Prob}(p) = f(p)\,dp.$$

Hence our formula can be written

$$\text{Prob}(p \mid k) = \frac{\binom{n}{k} p^k (1 - p)^{n-k} f(p)}{\int_0^1 \binom{n}{k} p^k (1 - p)^{n-k} f(p)\,dp}.$$

This is a special case of the classical *Bayes' formula*. It gives the "likelihood" that p is the true parameter, given that k was observed. The formula depends on the "prior belief," represented by the density function $f(p)$.

Prob $(p \mid k)$ can therefore be taken to be a distribution representing our belief about p when we combine the statistical experience and our prior belief.

14.6. If we know nothing about p except that k claims occurred in a sample of n, it may seem natural to assume $f(p) = 1$. This will reduce our formula to

$$\text{Prob}(p \mid k) = \frac{p^k (1 - p)^{n-k}}{\int_0^1 p^k (1 - p)^{n-k}\,dp} = (n + 1)\binom{n}{k} p^k (1 - p)^{n-k}.$$

Subjective Probabilities

We can now apply this result to our original problem. We started in 14.2 with the equation

$$U(S + P) - p\{U(S + P) - U(S + P - Z)\} = U(S).$$

We then multiplied by $f(p)\,dp$ and integrated over p from 0 to 1, and obtained

$$U(S + P) - \bar{p}\{U(S + P) - U(S + P - Z)\} = U(S)$$

where

$$\bar{p} = \int_0^1 pf(p)\,dp.$$

In our present example we have to replace $f(p)$ by

$$\text{Prob}\,(p \mid k) = (n + 1)\binom{n}{k}p^k(1 - p)^{n-k}.$$

Substituting this, we obtain

$$\bar{p} = (k + 1)/(n + 2).$$

This is then the probability which we should use in our decision if we want to combine the experience obtained by observing a comparable portfolio and our prior belief that every value of p was equally likely.

We see that for $k = 0$, we should take $\bar{p} = 1/(n + 2)$. This means that unless n is very large, our prior belief will still carry some weight. The fact that no claim occurred in a sample of n insurance contracts does not lead us to make future decisions on the assumption that $\bar{p} = 0$.

14.7. We have so far found that our prior belief about a parameter—the probability p in our example—could be represented by a weight function which had all the properties of a probability distribution. If this belief is as in the example of 14.3, the weight function may be as indicated by Fig. 16.

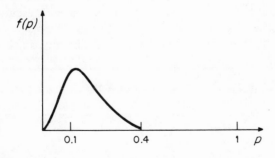

Figure 16

It is obvious that we usually will be interested only in the general shape of this function. We want a distribution function which can represent prior beliefs with sufficient approximation—or with as much precision as the decision-maker can express.

This is a parallel to what we found when we studied the utility functions which could represent the preferences of the decision-maker over a set of prospects. Since in practice we can only work with finite sets, there will usually be several functions which give an adequate representation. This means that we have some leeway in selecting the functions we use to represent prior beliefs and preferences. It is then natural to select functions which have a convenient mathematical form.

In the example we have discussed, we had to consider integrals of the form

$$\binom{n}{k} \int_0^1 p^k (1 - p)^{n-k} f(p) \, dx.$$

It is obvious that these integrals will become particularly easy to handle if $f(p)$ is a Beta-distribution, i.e., if

$$f(p) = \frac{(a + b + 1)!}{a! \, b!} p^a (1 - p)^b.$$

In the usual terminology [4], the Beta-distribution is called the *conjugate* of the binomial distribution.

We are therefore led to examine whether this class of distributions is sufficiently rich to represent all the prior beliefs we may want to study. For the mean of the distribution we find

$$\mu = \int_0^1 p f(p) \, dp = \frac{a + 1}{a + b + 2}$$

and for the variance:

$$\sigma^2 = \int_0^1 (p - \mu)^2 f(p) \, dp = \frac{(a + 1)(b + 1)}{(a + b + 2)^2 (a + b + 3)}.$$

From these expressions we see that if μ and σ are given, we can usually determine a and b. This means that if we get a sufficiently accurate description of our beliefs by specifying only the two first moments of the prior distribution, we can always find a Beta-distribution which meets our specifications.

14.8. Let us now assume that our prior belief can be represented by the distribution

$$f(p) = \frac{(a + b + 1)!}{a! \, b!} p^a (1 - p)^b.$$

Let us assume that when we have made up our mind in this way as to what we believe about the claim frequency, we learn that in a comparable portfolio of n contracts there were k claims. To make use of this new knowledge, we apply the formula in 14.5 and find

$$\text{Prob} \, (p \mid k) = \frac{p^{k+a}(1-p)^{n-k+b}}{\int_0^1 p^{k+a}(1-p)^{n-k+b} \, dp}$$

$$= \frac{(a+k)! \, (b+n-k)!}{(a+b+n-1)!} \, p^{k+a}(1-p)^{n-k+b}.$$

This gives the probability to be used for our decision

$$\bar{p} = \int_0^1 p \, \text{Prob} \, (p \mid k) \, dp = \frac{k+a+1}{n+a+b+1}.$$

As an illustration, let us assume that we feel certain that $p = p_0$. We must then have

$$p_0 = \mu = \frac{a+1}{a+b+2}$$

and

$$\sigma^2 = \frac{(a+1)(b+1)}{(a+b+2)^2(a+b+3)} = 0.$$

It is obvious that in this degenerate case both a and b must be infinite. From the first condition we obtain

$$p_0 = \lim \frac{\dfrac{a}{b} + \dfrac{1}{b}}{\dfrac{a}{b} + 1 + \dfrac{2}{b}} = \frac{\lim \left(\dfrac{a}{b}\right)}{\lim \left(\dfrac{a}{b}\right) + 1} = \lim \frac{a}{a+b}$$

or

$$\lim \left(\frac{a}{b}\right) = \frac{p_0}{1 - p_0}.$$

From the expression for $\bar{p}$ we find

$$\bar{p} = \lim \frac{\dfrac{k}{b} + \dfrac{a}{b} + \dfrac{1}{b}}{\dfrac{n}{b} + \dfrac{a}{b} + 1 + \dfrac{1}{b}} = \frac{\lim \left(\dfrac{a}{b}\right)}{\lim \left(\dfrac{a}{b}\right) + 1} = p_0.$$

This expresses the obvious. If we are certain that $p = p_0$, we will make our decision accordingly, no matter what experimental evidence should become available. The fact that k claims occurred in a portfolio of n contracts does not influence our decision, because our prior belief is as unshakable as if it was based on an infinite number of observations.

14.9. The example in the last paragraph indicates an attractive intuitive interpretation of the Beta-distribution. If our prior belief about the parameter p can be represented by the distribution

$$f(p) = \frac{(a+b+1)!}{a! \, b!} \, p^a(1-p)^b,$$

this belief is as if all we knew was that a claims were made in a portfolio of $a + b$ insurance contracts. This statement should have some intuitive meaning, at least to a statistician.

To look at the problem from another angle, let us assume that a person states that he believes there are 10 white and 90 black balls in an urn known to contain 100 balls. If he admits that he is not quite certain, we can ask him to formulate his uncertainty by tracing a curve of the type indicated by Fig. 16. It is very likely that our person will give up after a few attempts to draw a curve which gives an adequate representation of his state of mind. It may, however, occur to him that he can give an adequate description of his feeling of uncertainty in the following way:

If he had been allowed to draw a ball from the urn 50 times—putting the ball back after each drawing—and this had resulted in 5 white balls being drawn, he would believe that the urn contained about 10% white balls, without being quite certain. In fact, he would feel just as uncertain as he feels with regard to his original statement about prior belief.

More generally it should be possible to describe an experiment and specify an outcome of the experiment which would lead to a belief of the same strength as the prior belief which the decision-maker tries to describe. Prior belief can, as we have seen earlier, be represented by a weight function $f(p)$, defined over the domain of the uncertain parameter. It seems, however, more convenient to represent such beliefs by specifying an outcome of a well-defined experiment—both when the decision-maker tries to make up his own mind, and when he tries to communicate his beliefs to another person.

14.10. The ideas which we have outlined in the preceding section lead to the so-called *Bayesian approach to statistics*. It is usually assumed that this approach was discovered fairly recently—that it really originated with the book by Savage [5], already mentioned. These same ideas have, however, been discussed for more than 50 years—often in obscure language—by American actuaries, under the name of *credibility theory*. This theory was founded by Whitney [6], and has been developed by Perryman [3], Bailey [1], and others, without much contact with the mainstream of statistical theory. A recent paper by Mayerson [2] presents practically the whole credibility theory in modern statistical terminology.

To illustrate the application of the theory, let us consider an insurance company which has to quote a premium for an insurance contract of the simple type considered in our example. Let us assume that the company, when making this decision, can draw on two types of information:

(i) Statistical information about comparable contracts—for instance, that there were k claims in a portfolio of n identical or similar contracts.

(ii) Other relevant information—for instance, statistical observations of portfolios of contracts which are not quite comparable to the contract in question.

If there is sufficient statistical information, i.e., if n is large enough, the company will not consider the other information. The company will act as if it was certain that $p = k/n$. In this case, the actuaries will say that the statistical experience carries "100% credibility," but they will usually be embarrassed if they are asked to justify this statement.

When the statistical experience is insufficient, the company may use other relevant information. However, this is not possible unless different pieces of information can be made commensurable. This leads us to determine the statistical evidence which is equivalent to the other relevant information we want to use.

Theoretically an insurance company should bring into consideration additional information until the equivalent statistical experience carries 100% credibility. How this should be done is a difficult problem, which is far from being satisfactorily solved in the existing theory. The statistical experience of an insurance company which has written 1 million fire-insurance contracts in New York State obviously contains information which may be of value to a company which writes fire insurance in California. Intuitively we may feel that this information is worth less than information obtained from the statistical experience of a similar company operating in California. It is, however, not easy to give a precise formulation to such feelings. Will 1 million observations from New York be equivalent to 800,000 observations from California?

If an insurance company operates with a system of premium rates derived from statistical experience which carries 100% credibility, good or bad underwriting results will be explained as caused by random fluctuations. These results will not induce the company to change its premium rates.

In practice an insurance company will not usually assign 100% credibility to the statistical experience which constitutes the foundation of its premium rates. The most obvious reason for this cautious attitude is that the basic probabilities may change with time.

In this situation the company will accumulate new statistical experience as time goes by, and this new information may lead the company to adjust its premium rates. How much the rates should be changed will depend on the credibility carried by the initial statistical experience. This question can be the subject of heated discussions between company representatives and State Insurance Commissioners.

14.11. We shall now indicate how the ideas of the preceding sections can be generalized. Let us assume that the amount of claims x which will be made under a portfolio of insurance contracts is a stochastic variable with distribution $G(x, \alpha)$, where α is a vector of the parameters of the distribution function.

For the sake of simplicity, we shall assume that the distribution is continuous, and that a density function exists

$$g(x, \alpha) = \frac{\partial G(x, \alpha)}{\partial x}.$$

If the company has to quote a premium for this portfolio, it will compute the expected utility

$$\int_0^\infty U(S + P - x)g(x, \alpha)\, dx.$$

If there is some further uncertainty about the parameter α, expressed by a prior distribution $f(\alpha)$, we have to carry out another integration:

$$\int_A \left\{ \int_0^\infty U(S + P - x)g(x, \alpha)\, dx \right\} f(\alpha)\, d\alpha$$

where A is the domain of α. However, the latter integral is obviously equal to

$$\int_0^\infty U(S + P - x) \left\{ \int_A g(x, \alpha)f(\alpha)\, d\alpha \right\} dx.$$

The inner integral

$$h(x) = \int_A g(x, \alpha)f(\alpha)\, d\alpha$$

can, of course, be interpreted as a probability distribution. The expected utility is then

$$\int_A U(S + P - x)h(x)\, dx.$$

The whole reasoning about uncertainty over the value of the parameters means only that we replace the original distribution $g(x, \alpha)$ by $h(x)$. This means that our procedure has some practical value only if we know—or have good reasons to believe—that claim payments really are generated by a distribution of the form $g(x, \alpha)$. If this function is determined by fitting a curve to some data, i.e., by estimating the parameters α from these data, we don't gain anything by introducing a function $f(\alpha)$ to represent our subjective uncertainty about the estimates.

14.12. Let us now return to the expression

$$\int_0^\infty U(S + P - x) \left\{ \int_A g(x, \alpha)f(\alpha)\, d\alpha \right\} dx.$$

In this expression

  (i) The prior distribution $f(\alpha)$ represents what we *believe*.
  (ii) The utility function $U(x)$ represents what we *want*.
  (iii) The distribution $g(x, \alpha)$ represents what we *know*.

These three elements should all be considered in a rational decision, and in an analysis of the problem they should be separated.

In practice it may, however, be difficult to separate what we believe from what we know. In our simple example the place of the distribution $g(x, \alpha)$ was taken by the binomial

$$\binom{n}{k} p^k (1 - p)^{n-k}$$

where p is the unknown parameter. This binomial distribution rests on the assumptions

   (i) The probability of a claim is the same under all n contracts.

  (ii) The probability of a claim under an arbitrary contract is independent of whether claims have been made under any of the $n - 1$ other contracts.

If we *know* that these assumptions are true, there is no problem. If, however, these assumptions just represent our beliefs, or are accepted as working hypotheses, they should be included in $f(\alpha)$ and not in $g(x, \alpha)$. This means that the separation of the different elements, which seems essential to a rational analysis of the decision problem, is by its very nature arbitrary. This again means that in a preliminary study only the general shape of the functions $f(\alpha)$, $g(x, \alpha)$ and $U(x)$ is significant and that we should feel free to choose functions which are easy to manipulate mathematically.

REFERENCES

[1] Bailey, A. L.: "A Generalized Theory of Credibility," *Proceedings of the Casualty Actuarial Society*, Vol. XXXII (1945).

[2] Mayerson, A. L.: "A Bayesian View of Credibility," *Proceedings of the Casualty Actuarial Society*, Vol. LI (1964), pp. 85–104.

[3] Perryman, F. S.: "Some Notes on Credibility," *Proceedings of the Casualty Actuarial Society*, Vol. XIX (1932).

[4] Raiffa, H. and R. Schlaifer: *Applied Statistical Decision Theory*, Graduate School of Business Administration, Harvard University, 1961.

[5] Savage, L. J.: *The Foundation of Statistics*, Wiley, 1954.

[6] Whitney, A. W.: "The Theory of Experience Rating," *Proceedings of the Casualty Actuarial Society*, Vol. IV (1918).

Chapter XV

Group Decisions

15.1. In the preceding chapters we have generally considered our decision-maker as a single person who behaved in accordance with rules which could be accepted as "rational." In real life important economic decisions are often made, not by individuals, but by *groups* of persons who act as one single decision-maker. It is, therefore, of obvious interest to study the rules which govern the decisions of such groups—particularly if members of the group have conflicting interests so that the decision has to emerge as a compromise. In Chapter XIII we mentioned that such conflicts of interest might exist between the executives and the owners (share-holders) of a firm. As a result of such conflicts, some decisions made by the firm may seem strange to an outside observer, who may easily conclude that the firm does not act as a rational decision-maker.

The theory of group decisions is a vast subject, which probably belongs more to psychology and sociology than to economics. We shall, however, discuss it very briefly for the following two reasons:

(i) A decision made by a group can be considered as the outcome of a game—played by the members of the group. If they use mixed strategies, the outcome will be uncertain. Hence the fact that many decisions are made by groups may be an important source of uncertainty in the economy.

(ii) We have previously argued that there are certain basic rules which any rational person should observe in his decision-making. We shall see that a group of rational persons may easily violate these rules. This should make the reader a little tolerant when in real life he observes business decisions which appear irrational.

15.2. Most of the results we shall discuss are due to Arrow [1], and we shall begin by considering his introductory example, usually referred to as "the voting paradox."

Let us consider a group of three persons, who must select one of the three "candidates" *A*, *B*, or *C*. We can interpret the model as the problem of three brothers who have to hire a manager to run the family firm, or as an investment club, where the members debate which of the three securities they should buy.

Let us assume that the preference orderings of the three group members are

Person 1: *A, B, C*
Person 2: *C, A, B*
Person 3: *B, C, A.*

Let us next assume that the members of the group agree that the candidates are to be chosen by a majority vote. It is then easy to see the following:

One majority of the group (1 and 2) prefers *A* to *B*.
Another majority (1 and 3) prefers *B* to *C*.
A third majority (2 and 3) prefers *C* to *A*.

It thus seems that if the decision of the group is to be taken by majority vote, it must in some way depend on how the problem is presented, or on the voting procedure. This will hardly surprise an experienced politician or a board chairman, who knows that by "rigging" the agenda and the procedure he can obtain the decision he wants from a committee—within reason.

Even if this result is not surprising, it is nevertheless a little disturbing. Our starting point in Chapter III was that a rational person has a preference ordering over the set of objects (prospects) from which a choice has to be made. This seemed a very innocent assumption, really just implying that a rational decision-maker knows what he wants, and we have assumed that it holds for each member of the group.

A complete preference ordering is necessarily *transitive*. If the ordering is established by the relation, "is preferred to," the two statements

A is preferred to *B*.
B is preferred to *C*.

imply that

A is preferred to *C*.

Our example shows, however, that the implication does not hold for a group which makes its choice by majority vote. The group will express "circular" preferences when presented with a sequence of choices between two objects. This means that the group does not behave in a "rational" manner when judged by the standard which seemed appropriate for an individual decision-maker. Hence we are led to seek a meaningful definition of rational group decisions. Before we attack this problem, we shall make a little digression into psychology, and show that a situation similar to the "voting paradox" may occur also when we consider an individual decision-maker.

15.3. Let us consider a man—say an engineer—who has the choice of three jobs, one in New York (*N*), one in Paris (*P*), and one in Brasilia (*B*). Let us first assume that he looks at his problem as a professional man.

From this point of view there is no doubt that the job in Brasilia is the most attractive. The job will put him in charge of large construction works and will offer an exciting professional challenge. It is equally clear that the job in New York should be ranked second, since this will keep our engineer in the head office of a construction firm with world-wide operations. Paris then has to take last place in the ranking.

Let us next suppose that our engineer looks at his problem as the father of a family. From this point of view it is obvious that New York must be ranked first, since that city offers a wide choice of schools for his children. It is equally obvious that Paris must be ranked second, since that city also has excellent schools for English-speaking youngsters. Brasilia must clearly be ranked last, since moving to that city may mean a serious interruption in the education of the children.

As a third possibility, our engineer may look at the recreational possibilities offered by the three jobs. He may then rank Paris first, for reasons too obvious to be stated. Brasilia also offers possibilities of exciting recreation, so that New York has to take third place.

The decision problem can then be condensed into the following table:

Professional man: B, N, P
Family father: N, P, B
Recreations: P, B, N.

This table has exactly the same pattern as the table in 15.2, and our decision-making engineer appears as a split personality who is really a group of three persons with conflicting preferences.

We may well imagine that his actual decision will depend on his mood at the moment of decision. If he is tired and fed up with work in general, he may look only at the recreational possibilities and choose Paris. If, however, his weariness is due mainly to arguments with the children over their school reports, he may decide that his duties as father of a family should take priority, and will choose New York. If, on the other hand, our decision-maker feels energetic, and is craving for new professional challenges, he will obviously choose Brasilia.

This example shows that rationality is in a sense a relative concept. This is quite natural, since we have admitted that a decision rule must contain some subjective elements. To a psychologist it will then seem obvious that these subjective elements must depend on the mood of the decision-maker at the decisive moment.

These considerations will obviously apply also to group decisions. If, for instance, a decision on behalf of a firm is made by a committee made up of the managers of finance, production, and sales, we will get exactly the situation which we have discussed above. The decision actually made may well be determined by the fact that the sales manager was in a particularly boisterous mood when the committee met and that he swept his two colleagues with him. It is not a new idea that the ulcers of executives have some influence on the national economy.

15.4. Let us now return to our main problem and try to formulate the rules for group decisions in a more general way than simple majority voting.

In the example considered in 15.2, there are six possible preference orderings of the three candidates, and these can be represented by six row vectors. Let R be the set of these orderings $R_1, R_2, \ldots, R_6$, i.e.,

$$R_1 = \{A, B, C\}$$
$$R_2 = \{A, C, B\}$$
$$R_3 = \{B, A, C\}$$
$$R_4 = \{B, C, A\}$$
$$R_5 = \{C, A, B\}$$
$$R_6 = \{C, B, A\}.$$

Let us now assume that two persons have the preference orderings R_1 and R_4 respectively. The individual preferences of this group of two persons can then be completely described by the matrix

$$\begin{bmatrix} A & B & C \\ B & C & A \end{bmatrix}$$

A rule for group decisions must tell us how the preference ordering of the group is derived from the preference orderings of the members. The ordering of the group must necessarily be one of the six elements of R. The rule must then be a function which, for each group matrix, gives us an element of the set R, i.e., a function with R as its range, and the Cartesian product $R \times R$ as its domain. A function of this kind can obviously be defined for any number of candidates, and for groups with any number of members. Our problem is now to lay down some general conditions which this function should fulfill in order to give a reasonable rule for deriving the preference ordering of the group.

15.5. The problem we have outlined in the preceding paragraph was first formulated in this general manner by Arrow [1], and we shall consider the five conditions he laid down in the first edition of his book. The conditions can be formulated in many different ways, and these alternatives are discussed by Luce and Raiffa ([2], Chap. 14) and by Arrow himself in the second edition of his book.

Condition 1 simply states that the function giving the group preference exists when there are at least three candidates.

Condition 2 requires *positive association* between individual preferences and the preference ordering of the group.

To illustrate, let us assume that the preference ordering of the two-person group in the preceding paragraph is $\{B, A, C\}$. We shall express this as follows:

$$\begin{bmatrix} A & B & C \\ B & C & A \end{bmatrix} \longrightarrow \{B, A, C\}.$$

Suppose now that person 2 changes his preference ordering to $\{B, A, C\}$; i.e., he decides to rank A before C, but still after B. The condition states that we then have

$$\begin{bmatrix} A & B & C \\ B & A & C \end{bmatrix} \longrightarrow \{B, A, C\};$$

i.e., if A is ranked higher in the preference ordering of an individual, it cannot be given a lower ranking by the group.

Condition 3 states that the group ordering of the remaining candidates shall not change if some candidates are removed. This means for instance that

$$\begin{bmatrix} A & B & C \\ B & C & A \end{bmatrix} \longrightarrow \{B, A, C\}$$

implies

$$\begin{bmatrix} A & B \\ B & A \end{bmatrix} \longrightarrow \{B, A\}$$

$$\begin{bmatrix} A & C \\ C & A \end{bmatrix} \longrightarrow \{A, C\}$$

and

$$\begin{bmatrix} B & C \\ B & C \end{bmatrix} \longrightarrow \{B, C\}.$$

This is the principle of *independence of irrelevant alternatives*, which we have encountered earlier—as Condition (iii) in 10.11.

Condition 4 states that the preference ordering of the group must not be *imposed*; i.e., it must not be independent of individual preferences.

Condition 5 requires that there shall be *no dictatorship*; i.e., the preference ordering of the group shall not be identical to that of a particular individual, regardless of the preferences held by the other members.

These five conditions appear very innocent, and they may seem very reasonable as the "minimum requirements" for a rule as to how a group should make its decision with due respect to the wishes of individual members. Actually the conditions are too strong.

Arrow's *Impossibility Theorem* states that a rule which satisfies Conditions 1, 2, and 3 is either imposed or dictatorial.

15.6. The five conditions can, as mentioned earlier, be given in several different forms. For instance, it is easy from conditions 1, 2, 3, and 4 to prove Condition *P*:

Condition P: If all individuals have the same preference ordering, this will also be the preference ordering of the group. This means that

$$\begin{bmatrix} A & C & B \\ A & C & B \\ A & C & B \end{bmatrix} \longrightarrow \{A, C, B\}.$$

We could call this condition the principle of *unanimity*, or of *Pareto optimality*. It seems in a way simpler and more "obviously acceptable" than some of the other conditions, and it is also possible to prove the Impossibility Theorem from conditions 1, 3, 5, and *P*.

To illustrate the essentially combinatorial nature of the theorem, let us consider a group of two persons, and two candidates *A* and *B*. From condition *P* it follows that

$$\begin{bmatrix} A & B \\ A & B \end{bmatrix} \longrightarrow \{A, B\}$$

and that

$$\begin{bmatrix} A & B \\ B & A \end{bmatrix} \longrightarrow \{A, B\} \quad \text{or} \quad \{B, A\}.$$

In the former case person 1 is a dictator. In the latter case person 2 is the dictator, since

$$\begin{bmatrix} B & A \\ B & A \end{bmatrix} \longrightarrow \{B, A\}.$$

Let us now take the first of these two cases, and see if we can avoid the need for dictatorship by bringing in more candidates. It is easy to see that

$$\begin{bmatrix} A & B & C \\ B & A & C \end{bmatrix} \longrightarrow \{A, B, C\}.$$

Both persons rank *C* last, and the existence of *C* cannot change the group's ordering of *A* and *B*.

It is also clear that

$$\begin{bmatrix} A & B & C \\ B & C & A \end{bmatrix} \longrightarrow \{A, B, C\}$$

since both persons agree that *C* shall be ranked after *B*. It is also easy to see that

$$\begin{bmatrix} A & B & C \\ C & B & A \end{bmatrix} \longrightarrow \{A, B, C\}, \{A, C, B\}, \text{ or } \{C, A, B\}.$$

In the first of these three cases, it is clear that person 1 is a dictator, since his ordering prevails, no matter what are the preference of person 2. To prove this completely, we should show that the group ordering is $\{A, B, C\}$ also when person 2 has the ordering (C, A, B), (A, C, B), and (A, B, C), i.e., when he agrees that A should be ranked before B. This should, however, be obvious.

The two last cases seem to hold some promise of avoiding dictatorship. Let us therefore examine the second case, i.e.,

$$\begin{bmatrix} A & B & C \\ C & B & A \end{bmatrix} \longrightarrow \{A, C, B\}.$$

Applying condition 3, we find

$$\begin{bmatrix} A & B \\ B & A \end{bmatrix} \longrightarrow \{A, B\}$$

$$\begin{bmatrix} A & C \\ C & A \end{bmatrix} \longrightarrow \{A, C\}$$

$$\begin{bmatrix} B & C \\ C & B \end{bmatrix} \longrightarrow \{C, B\}.$$

This means that

Person 1 decides the group ordering of the pairs (A, B) and (A, C).
Person 2 decides the group ordering of the pair (B, C).

This seems promising, since it is clear that the ban on dictatorship implies that no person can decide the group ordering of all pairs—and vice versa.

In our example, person 1 could use his power to decide that A should be before B and C in the group ordering, and then let person 2 decide the ordering of B and C. This seems to be a nice, almost democratic arrangement, but it is an illusion. If person 1 decides that A shall come after B but before C in the group ordering, he has also decided that B shall come before C and has thus made the right of person 2 worthless. The alternatives are then: Dictatorship by person 1, or a circular ordering for the group.

15.7. In the preceding section we have tried to bring out the main ideas behind Arrow's Impossibility Theorem. We shall now give a brief outline of the general proof.

    (i) From condition 3 and P it follows that if the group prefers A to B, this preference must be held by at least one member. We shall call the set of members who rank A before B the *support* of the ordered pair (A, B) in the group ordering.

    (ii) Assume that the group prefers A to B, and let (A, B) be a pair in the group ordering such that no other pair has a smaller support.

Let the support of (A, B) consist of

> Person 1 with the preference ordering (A, B, C).
> A set V of members with the ordering (C, A, B).

Let U be the set of members who do not support (A, B). We assume that they all have the preference ordering (B, C, A).

The group must then prefer B to C. If it did not, the ordered pair (C, B) would be in the group ordering with the support V; i.e., it would have a smaller support than (A, B), contrary to our assumption. The preference ordering of the group must then be (A, B, C).

(iii) The support of (A, C) consists of only player 1. This will be smaller than the support of (A, B) unless the set V is empty.

(iv) We have now proved that person 1 alone decides the group ordering of any pair of the type (A, C). In order to be a dictator, he must also be able to decide the ordering of pairs of the type (B, A) and (B, C).

Let us first consider the preference orderings

> Person 1: (B, A, C)
> Set U: (C, B, A).

From condition P it follows that the group will rank B before A, which again is ranked before C. Hence person 1 decides the group ordering of (B, C).

Let us next consider the orderings

> Person 1: (B, C, A)
> Set U: (C, A, B).

Person 1 decides that the group shall rank B before C, and C is, by condition P, ranked before A. Hence the group ranks B before A; i.e., the ordering of person 1 is decisive. It then follows that person 1 is a dictator, and this contradicts condition 5.

15.8. It is obvious that Arrow's theorem has far-reaching implications for the social sciences. These implications have been discussed in considerable detail by Luce and Raiffa [2] and other authors. We shall, therefore, only make a few remarks to connect the theorem with the problems discussed in earlier chapters.

If there is a high degree of consensus among the members of a group, it is clear that the group can make its decisions by simple majority voting, without running into difficulties. This means that in practice we can often ignore the theoretical impossibility of a perfect, democratic decision procedure.

In Arrow's model we consider preferences without making any assumptions about their strength. The following arrangement implies that person 1 is a dictator:

$$\begin{bmatrix} A & B & C \\ C & B & A \\ C & B & A \end{bmatrix} \longrightarrow \{A, B, C\}.$$

The arrangement may be considered as quite reasonable if persons 2 and 3 are practically indifferent about the candidates, and if person 1 thinks the election of C will be a disaster. If we want to build a theory on the basis indicated by this example, we will have to introduce *interpersonal comparability of utility*. This is a tempting emergency exit, which one day may be necessary to save the social sciences from logical inconsistency. We have, however, so far refused to take this way out, and we shall not do so now.

15.9. In order to develop a satisfactory theory for rational group decisions, we must evidently relax or drop some of the five conditions discussed in 15.5. In the context of this book it seems most natural to give up condition 1, which implies that individual preferences completely determine the preference ordering of the group. In our discussion of game theory we found that complete knowledge about the rules of the game and the objectives of the players would in general only make it possible to specify a probability distribution over the outcomes of the game. This should indicate that group decisions and group preferences can only be predicted in a probabilistic sense, even if we have full knowledge about individual preferences.

This conclusion is supported by observations from real life. If a decision problem leads to a genuine tie, the traditional solution is to draw lots. This is, however, not a unique solution, since one can draw lots in several different ways. As an illustration, let us again look at the situation considered in 15.2.

Once the three persons in the group realize that they are deadlocked, they may agree to leave the decision to chance, and they may consider the following three possibilities:

(i) They can select a candidate by drawing lots. This may be rejected as too unsophisicated a solution. It may also be non-optimal in a general case. If there is some consensus about the candidates, it does not seem right to give them all the same chance of being elected.

(ii) The members of the group can draw lots to select one among themselves to serve as dictator. This solution may be objectionable to good democrats, and it gives one person more power than necessary to break ties.

(iii) Our persons may select a chairman by drawing lots or—to be more realistic—agree that the chairmanship shall rotate. The chairman may be given the right to prepare the agenda, i.e., to determine how the group shall vote over the candidates. This will usually make it possible for the group to reach a decision—elect a candidate, or establish a preference ordering.

We shall not go deeper into these problems, since the problem of devising optimal voting systems is a subject in itself. To sum up the situation, let us again note that a group in general is unable to make a decision if the opinions of the members are sufficiently dispersed. To enable the group to make a decision in all cases, we may select one member at random, and give him the special powers necessary to break any ties which may occur. The problem is then to determine the minimum amount of power this member must have to make sure that there can be no ties. The solution may well be a constitutional monarch who exercises power only in situations which the constitution has not foreseen.

15.10. To close this chapter, let us consider a group of two persons who have to select one of two prospects, described by the probability distributions $f(x)$ and $g(x)$. If they both prefer $f(x)$ to $g(x)$, it seems obvious that the group should select $f(x)$, i.e.

$$\begin{bmatrix} f(x) & g(x) \\ f(x) & g(x) \end{bmatrix} \longrightarrow \{f(x), g(x)\}.$$

In order to establish a connection between individual preferences and group preferences, we must assume that the two persons have agreed upon some rule for dividing the return from the prospect selected by the group. For the sake of simplicity we shall assume that person 1 gets a fraction t_1 of the return, and that person 2 gets a fraction $t_2 = 1 - t_1$.

Let now $u_i(x)$ be the utility function which represents the preference ordering of person i. Our assumption about individual preferences implies

$$\sum u_i(x)f(x) > \sum u_i(x)g(x) \quad \text{for } i = 1, 2.$$

It is obvious that under certain conditions we can find values of t_1 and t_2 so that

$$\sum u_i(t_i x)g(x) > \sum u_i(t_i x)f(x).$$

This implies, however, that both persons agree that the group shall choose $g(x)$.

As an example, let us take $u_i(x) = x - a_i x^2$. The two inequalities then become

$$E_f - a_i E_f^2 - a_i V_f > E_g - a_i E_g^2 - a_i V_g$$

and

$$E_g - a_i t_i E_g^2 - a_i t_i V_g > E_f - a_i t_i E_f^2 - a_i t_i V_f.$$

From this we obtain the double inequality

$$t_i\{E_g{}^2 - E_f{}^2 + V_g - V_f\} < (1/a_i)(E_g - E_f) < E_g{}^2 - E_f{}^2 + V_g - V_f.$$

Here E_f, E_g, V_f, and V_g are the means and variances of the returns from the two prospects.

To obtain a numerical example, let us take

$$t_1 = t_2 = \tfrac{1}{2}$$
$$E_f = 1 \quad \text{and} \quad E_g = 2$$
$$V_f = 1 \quad \text{and} \quad V_g = 12.$$

The double inequality is then reduced to

$$1/14 < a_i < 2/14.$$

For the numerical values we have chosen, $f(x)$ and $g(x)$ can be taken to be the two binary prospects $(\tfrac{1}{2}, 2)$ and $(\tfrac{1}{4}, 8)$.

If the "coefficient of risk aversion" a_i satisfies the double inequality for both persons, they will both prefer $(\tfrac{1}{2}, 2)$ to $(\tfrac{1}{4}, 8)$. If, however, they have agreed that the profits they make as a group shall be split 50–50, the group will prefer $(\tfrac{1}{4}, 8)$ to $(\tfrac{1}{2}, 2)$. The reason for this apparent contradiction is that both persons prefer $(\tfrac{1}{4}, 4)$ to $(\tfrac{1}{2}, 1)$.

15.11. We stated earlier that it would be useful to discuss group decisions if it would make the reader more tolerant about behavior which may seem irrational. It is hoped that our examples will make it clear that very few behavioral assumptions can be taken as "obviously rational" in every possible context.

Assumptions which we found hard to reject as rational for a single decision-maker become not only dubious or unrealistic, but logically impossible when applied to groups.

We also found that an assumption such as "the principle of unanimity," which appears an absolute, minimal requirement of democratic procedures, had to be revised when we brought uncertainty into consideration.

The desire to be tolerant should not lead us to accept every possible assumption. When two assumptions contradict each other, it is clear that at least one of them must be rejected. It is, however, not always easy to see when a contradiction exists between some almost completely unrelated assumptions. The theory of group decisions offers some fine examples where we have to dig deep to uncover the contradiction.

REFERENCES

[1] Arrow, K. J.: *Social Choice and Individual Values*, Wiley, 1951, 2nd ed., 1963.
[2] Luce, R. D. and H. Raiffa: *Games and Decisions*, Wiley, 1957.

Index